New Texas

Kyle Huizinga

*"I've finished my war book now.

The next one I write is going to be fun.

This one is a failure, and had to be, since it was written by a pillar of salt.

It begins like this.

Listen."

-Kurt Vonnegut

Prelude – The Parents

They kiss -- one of thousands that had the same passion as the first.   Large pupils focus on everything and nothing before they clench, peering into the other's soul for their reflected love; a deep unrelenting love. They help each other with the tight suits, helmets getting locked into place while beating hearts race. Her man enters the vehicle first to be strapped to the seat. She took her place sitting on top of him getting strapped to her man. The last strap she undid to make their journey one that they both would love. The crew chief shaking his head with a slight grin, he knew that they would do what they wanted, and he still felt fear from these two even in their lightest most humanizing hour.

The checks were made before the countdown to launch. Most would be worried but they only thought of each other. She actually couldn't wait, they were close but it made her desire for him unbearable. Squirming on his lap she turned her head into his broad chest, felt his heart pound through muscular body. The countdown ended and the shaking, pressing acceleration began. Speed crushing her into him uncomfortably, but he would take much more. He would take anything from her; he would die if she wanted him to.

 Soon the flight leveled out, steadied. She felt for his legs and hips, felt for the clumsy leather obstacles to her goal. She wanted her man one last time, wanted his pleasure. He helped her with the suits feeling her firm thighs as he squeezed. Everything was so confined in the small space that was more like a capsule than a cockpit. She barely had room to be lifted onto her man -her perfect man- they were made for each other. She moaned as she let him inside, quivers as icy pleasure runs up her spine. She couldn't help but move, to feel her man. Always love before anger- pleasure before pain.

One hundred forty three minutes later she screams, more out of desire... of anger, he squeezes her with large hands and they were two separate beings. Cool sweat and shivers cooling where it could, hearts pounding steadily, she would miss him. And they both did not want to put the suits back in order, but they had a mission to do.
Their countries' missiles were en route to clear the way. Small deadly bodies screaming by sooner than expected, this was their entrance- they just yelled "open sesame"!

***

The dirt exploding beneath them as their descent was quick to the ground. Their countries bombs leveling the ground and leading the way. The U.E.O was the only manufacturer to make such an incredibly reactive and tensile material. They made a material that was out of this world astronomical, and many radioactive materials combined to form the most deadly of weapons. But the U.E.O could not even control it. Uranium mixed with Einsteinium, Iridium, and Plutonium among others, creating a terrible mix that nobody wanted to touch, a name nobody wanted to mention in something nobody should ever want to handle, because this material had already taken the lives of thousands in its creation.

This was the lover assassins' goal. To steal the damned material that couldn't be handled. Its combined atomic power enhanced as this compound was like a chunk of the core of the sun, manufactured by humans- for war. Its power would gain hierarchy in its destructive capability, but this material was locked away for its own good and for the good of the world.

***

Their aircraft slowed with a crushing multiplying G-force. They were already underground in the lair that held the dragon's gold, the craft landing softly in a tunnel next to the broken entrance and crumbling debris. Straps unlocked, gear in hand, and they dove out of the one seat. Hot skin made suits tight with sweat; they move down the long tunnel towards the increasing Geiger beeps. They turn a corner and he pulled her back as motion detecting mini-guns unleash a rain of fire. A second later and two EMP grenades were tossed around the bend. Loading penetrating rockets they wait for the burst of electromagnetic static that made a tingling sensation on their damp confined hair. They move and two penetrating rockets slam into the heavy leaden door- but it barely moved. They would have looked worried if not concealed by helmets; they reload and try again with greater effect as the heavy door bent inward. One more barrage and the doors flung open to expose a wave of heat and a spike of Geiger beeps which was now a constant flow. They were careful entering the room to browse for more automatic weaponry as they notice that the room seemed unfinished in haste to store the nasty substance. Dark was pierced by light creeping out of small metal storage containers in three cubic feet apiece. They move to carry one box at a time as even a small amount of friction on the container could be catastrophic. They get to eight but the informant said there should have been ten. He told her to keep loading the craft as he would look for the rest. He found them easily with the Geiger, down a small corridor from the storage and into another large room housing two half-built rockets from a discontinued project. He noticed the two empty containers and went to the rockets to open the payload. It took a minute to figure out the unlocking mechanism but it was done as white hot material unleashed a blinding radioactive energy. He took the time to put on dusty reinforced gloves and flexed thick muscle to lift the material into containers that were fortunately close. The last brick slid into place which immediately put off a shockwave of blistering heat. He closed the lid in a gasp as his muscular body felt terribly weak. He opened a pocket to expose an adrenaline shot laced with painkiller and slammed the shot into his thigh. Large muscles bulging as he carried one of the large packages all the way to the corridor. Internal decay nearly doubled him over as he realized that his lover was taking fire from the fast acting military. Most likely Special Forces, but he was already angry and ready for a fight.

Two small sub machineguns were produced as he crept forward to a better vantage point. His lover took cover behind the unloaded material which he realized was a good idea as laser sights waived around unwilling to shoot and 'ignite' the substance. They were moving dangerously close to his lover in a tight creeping line. He had one more rocket and he loaded the penetrator and fired on the line of soldiers. The missile swatted humans and he moved from his corner to feel the impact exposing himself to the sharp shooters who were bunkered down. Sighting through holo-scopes with bull-pup guns he moved from cover to take a couple shots which gave away the shooters position. His guns dispatched these trained killers as if they were not. A tenderized chest felt pummeled and hot, but the armor did its job and he moved to his lover. She had been shot more than once and in the shoulder in a place where the armor was soft which most likely had broken a bone or two. She still hugged him through this anguish, and they move to get the craft loaded while dumbfounded militaries scrambled to figure out who and what was going on.

***

They nearly finished their job and took off stuffy helmets to reveal bleeding eyes. Radioactive decay had unleashed its deadly fire. Heads twang with agony as she crumpled in sharp pain. He gave her a large dose of painkiller and adrenaline. She knew it was best as they move the last two bricks to the loaders entry and closed it in relief. The hot overactive energy was absorbed by the rocket's boosters and a sharp quick pulse lifts the rocket into the sky. The Superpowers of the world would be afraid of those who gain this power.

They fall back into the bunker of deaths shadow as the melting heat was felt in decaying bones. They always laughed at death as they were soldiers of a higher cause, enveloped with a heightened sense of love as their eyes shed tears of blood. Their embrace was not finite- at this moment it was not movable. They locked heart and soul as only the most doomed

could do in their last minute, could not help but kiss
and hug as they lay on the ground, their bodies a
waste of radioactive mess. They lit the small igniters
of the suits, lined with thermite liquid and they begin
to burn. Two bodies connected blossom into white
and orange -gray and blue- fire and ash. A swirling
hot cloud lifts into the night while wasted bombs
start to burst around wasted land.

Speed and power - 2109

Her mind and body moved in its naturally graceful nature. Mental focus was free as she lunged and struck. Her mental time machine slowed the crush of the ball, placing her feet, breath escaping her lungs. Her brother's wild powerful arch sent a challenge that she was proud of. She moved with quickness that her own sweat could not follow, her slight crouch flowing into a rocket that spun backwards for a backhanded strike. Alluana landed and moved for the challenge as it was already over. The ball hit Titan directly in his chest and she frowned when she noticed him smiling. 'He should have tried,' she thought, 'it was too easy.'

Titan knew that he could have moved but was frozen in awe for that fleeting second. He began congratulating his sister before the ball landed on the ground. A large smile spread across his gleaming teeth, but his sister seemed angry. It made him think of their parents. Titan rarely thought of them as they were but a fading shadow of a distant memory. A picture was always present as it hung on their wall- his mother and his father. They were always so big and strong in his past memories. but parents that were gone since they were five and they were killed in a flight. He never questioned that, but he questioned her angry look.

"You don't like your win, Luna?"

"I haven't won yet Titan, it's still a tie."

He thought back as he looked at the clock. Nearly an hour had passed and only four points. 'Damn' he thought, 'when would this game ever end?' He wanted to give her the win but knew it would just make her even angrier.

"Can't we play a tie breaker?"

"No we made the rules before we started, and we are going to finish it that way."

"So now what's the score?"

"We're up to 98-98, Ti. What, you tired?"

"No, just have to work tomorrow, that's all."

"Aw, how sad, I have to work tomorrow too."

"My work… I'm hungry, let's go eat." Titan knew he couldn't complain about work to his sister but it seemed like they were pushing him, and always with more strange tasks and with different bosses.

"No complaint there, I can always win on a different day. Only two more points."

'Playing to a hundred,' Titan thought. 'How did it take so long?' It was half a year ago that they started playing this stupid game- which was never ending. Titan did not like the rules so they decided to make their own. 'What a mistake,' Titan thought, as he now realized why the game had been played with sets for scoring. 'First to a hundred would be easy,' he remembered thinking. 'No sets would be less confusing and would end the game faster… a quick hundred.'

"Afraid I might win your game?" Luna asked with a huge grin on her face. She was incredibly adept at picking one's strings as they were in thought.

"Not a chance," Titan said, as he could not be the one to look weak at any time, but he felt as though the taste for this game had left his palate. He figured that Luna would have quit long ago. After all, he usually won at the smaller Ping-Pong tables in the bars. Maybe she was the one who didn't like the game, but now tennis was becoming her own. He did like the sport. 'Or was it too competitive?

Didn't his sister win already anyways?'

"Let's go get some sushi." Luna stated.

The other thing Titan didn't like- sushi.

"Okay, let's get sushi, and I will win, when we play next."

"Sure you will big guy, I'll meet you outside."

Changing in the locker room just compiled his aggravation. 'Didn't she win already? She must still want to play? Why wouldn't she have told him if she had made the 100?' He hated to lose but at the same time wanted to congratulate his own sister who still wanted to kick his ass. And now he did not even like the game which he started. Funny how things work, somehow a fun game turned into a fanatical competition which only made him question his own sister's motives. 'Why would she not want to win? Didn't she win?'
The smiles that he got and the walk to the restaurant only made things worse as he wanted to ask what the actual score was. He was going to ask, but she made him forget with her first question on their stroll to the restaurant.

"Are you ever going to get a girlfriend?"

The question put him off as it was the last thing that he was thinking about at the moment.

"Maybe, are you ever going to get a boyfriend?"

"Titan, I asked you first."

Why did she never answer his questions? Or was it only his imagination as they had been speaking more of their past and future endeavors, but she still avoided most human contact. 'Funny,' Titan thought as he looked upwards towards the giant Metro buildings they had started walking between, 'she wants to know all of the different peoples of the world and yet she stays away from human intimacy… and maybe that's her plan?' He asked himself as he noticed her looking at him and quickly formed the words that he needed to say.

"Like I said before… eventually I will meet somebody."

"You have plenty to choose from, and you finally got over your puppy love, that one girl."

"Well, you're coming with me to the nightclubs next week, right?"

"Yes," Luna said quickly on a dull note as Titan knew it was the place that she did not want to go. And what she asked next amazed him even more about her natural ability to pick ones mind.

"Have you thought of our parents lately?"

"Yes… sometimes I think I dream about them."

"I dream of them too sometimes, but not very often." Luna said, as she thought to herself that this was probably the beginning of their adulthood with adult conversation, even though it was about dreams. She sometimes thought to herself about being more mature. And when you grow up all you can think about is being young and free.

"It always seemed like they were both angry about something, or like they traded anger- one after the other. And I only dream of mom being angry," Titan explained as he thought of the strange compiling dreams that tended to be about current events.

"That's weird, I only dream of dad being angry." She smiled and they both laughed a little before she asked, "I wonder what they were always mad about? They traveled and got to see many places and different people."

"YEAH, and they said it was cool until they had to work, and they never once took us!" Titan burst out, as if it had been brewing on his mind for too long and finally was said.

Luna couldn't help but smile as she replied.
"But they were always fighting for our rights they said, and that this was the greatest country in the archaic world."

"I don't see why, America and Mexico aren't bad neighbors, and they have fun amusement parks."

"They took us to the border countries once each, but I still want to see different continents and exotic things."

"That's what I meant, they never took us across to any other continents, and I barely remember the border trips, but now I don't care because the Acropolis has all of the different landscapes you would ever want to go."

"Man-made, it's where everybody goes, that's a joke, Titan."

"It's that or standing in a river because the rest of the world is already explored."

Fury burned in Luna's eyes as she replied,
"The world can never be fully explored, Titan. And only the weak would not want to."

Titan was taken aback for a second as he carelessly hit a defensive chord in his sister. 'Stupid' he thought, as he wanted the night to go smoothly.

"You're right; I try to find something new in Terra everyday. And maybe I will find a girlfriend that you so desperately want me to have."

Titan was actually relieved to open the door to the Sushi Bar and hoped that they wouldn't speak of this anymore. His mind however, would not stop thinking of what she had said. *Man-made... that's a joke, Titan.* The winding concrete path that they took seemed so beautiful and full of life as many plants veiled most of the man-made structures with organic elegance. Even the towering structures were spires of green, at least where no solar panels were present, and Titan somewhat questioned this motive.

Luna also was found in quiet thought; they sat at the bar to order soup and sake. She loved her country, but wanted to see the sites where ancient people roamed, and she wanted to climb the same mountains even if they had been explored. She glanced at her brother and found him in his own thought-provoking state and she wanted him to be happy. She wanted him to have a woman, but it was also something that she did not want Titan to have for some reason. 'Jealousy, for my own brother, and what a childlike reaction,' Luna thought, as she took a sip of sake and finally replied,

"Someday I might meet someone too, but you can still live with me."

"Yeah, I need the big bedroom though, I like my women big."

"I see a lot of American women at the club we're going to. Your chances should be good." Luna smiled even though most Americans she had seen were not too large.

"Big hearts made for big minds, and you know I like them smart" Titan remarked.

"Well you lucked out because I'm your sister." It slipped out of her mouth and a nervous beat twitched her eyes as she followed with, "And will make sure you find the one who might challenge me."

"Yeah, you know that won't happen, but I know you will try." Titan said bracing himself against thinking of losing that game.
"So where are you going first on your trip?" Titan asked.

"Probably Amsterdam, the Hague is still the place. Or maybe Tokyo, because it's on the way," Luna replied quickly as she was glad that the subject had been changed.

"Maybe I can meet you somewhere half way through your trip, but you know how they like to work me."

"We will take a vacation Ti, it's just a matter of time."

They finished the meal in playful laughter as Titan and Luna could always find the brighter side to any situation. Something that their parents had taught them from the beginning as even silent meditation could create worlds of imagination in a land far away. Entertainment was always the biggest priority in their country that needed hard work for the quality of life. And the people were more than willing to pay their share as it seemed that nobody wanted to leave this paradise. A man-made utopia had been created as a perfect home for many, including the twins.

***

Titan was walking through the labyrinthine gardens of his home. Water flowed from fountains pumped to the top of the massive multi-tiered pyramids. The flowing water was serene, consistently whispering through its prismatic distortion. Children were laughing through the echoing water but Titan could not see anybody. Titan was now floating in the middle of one of these pools from which water flowed into and out of. He was standing on a giant water lily trimming vines teaming with flowers and fruit. They were moving up and down as Titan was also moving up and down on the slight current which started to funnel into a bubbling stream. This stream became a torrent but Titan was not afraid as he saw people ahead ready to help him out. They did, and they were a crowd but the only two who stood out were his parents who wore their large necklace. They didn't look like his parents, they were too young and they were naked. Even though they were naked Titan could not focus on anything his mind did not want to look at. They were smooth like dolls, and his mother took his hand and they went into the Abbey. The churchyard was filled with naked people who all had no privates- it was very foggy. His mother's translator necklace was speaking to him, although her lips did not move. It was whispering something comforting something that she last said, something as simple as, "It's okay, It's going to be okay- I love you." Titan felt sad but he realized that even the preacher was bowing down to his mother. She was grinning, a beautiful almost crazy smile, and Titan looked into the swirling clouds above and when he brought his eyes back to his mother she was strangling the preacher. The preachers face was turning black and the bodies around them were turning black. They were writhing on the ground as his mother calmly said to her son, "It's okay if they all die." She had black eyes and black lips and she was going to kiss her son, he was going to kiss her back but he woke in a twisted mass of sheets.

--

Alluana was sitting in class. The man teaching had a large skin-toned necklace wrapped tight to his neck. His veins were bulging around the necklace as he spoke. It was form fitting, a piece of soft rectangular padding and metal strapped around the throat and it looked uncomfortable. The man seemed like he did not even notice its presence as he addressed the class in one language, then another, and another. Children were scrambling to write one line after the next. Her father's eyes locked onto Alluana's and he asked her, "Why are you not writing?"

She replied defiantly, "Because I already know!" Her father smiled and nodded approvingly. He took her hand and they walked onto the balcony. She looked from east to west and north to south and could see her entire country as she was on top of the highest pillar. There were cranes hoisting building materials into massive cylindrical buildings. Beyond were glistening reservoirs surrounding glistening green pyramids. She looked north into buildings that looked just as fresh. Shining in their own light really, even though the build was more like a dark Gothic Victorian mixture. She moved east to the only buildings that were square, although they didn't look it. They were massive domed hangars creating a rolling mountainous desert of buildings, as if they said, 'just try to walk over me and you will fail.' Her view brought her south again, a mixture of food production and water treatment was in this district but it mixed with the residential buildings and they all looked mostly the same. She felt a sense of higher being, like she was the pinnacle of the world and the world revolved around her. Her presence and her being giving off as much energy as was given, and she knew that there was no place better than where she was standing, no place more compromising or forgiving. Her crest broke and she felt sad as if there was nothing else that she could do. Her father put his hand on her shoulder, it was hard and calloused. He said, "Let me show you Ana." He took her hand and they started to fly. She was cruising between galaxies in a spaceship with her father. Heavens paintbrush was dark but its matter was undeniably beautiful. The blackness held waves of greens and blues mixed with a fantastic, brilliant, silver and gold. Alluana remembered that only her family ever called her Ana, everybody else had started to use Luna. She felt elated, she absolutely loved the rough edges of her father, she had cried for weeks but he was with her now. They weren't even in the spaceship anymore they were floating, hand in hand. He turned to face her floating backwards he told her that she can't know everything. He told her it would be okay even though his lips didn't move. He pushed her back as the blackness of the funnel folded him into nothing. She was alone and she woke to tears from a dream.

Meltdown – 2109

A beautiful frown deepens.

"What's wrong, Luna?" Titan asks.

"Nothing … it's just- I've never heard any really

good news from the outside world, Ti."

"You are too sweet sis; there are bad people even in Terra."

"Yeah, when they get too drunk at the night clubs, and you know how they calm down when we show up."

"There was that guy who hit his kids, Luna."

"Damn it, Ti. You know what I mean, and you know the punishment that drunker asshole got from his judge."

Ti knew she wasn't to be messed with when she used words like that, and he tried to stay silent.
"Good neighbor." Ti whispered.

"THEN WHY DON'T YOU REALIZE THE REST OF THE WORLD COULD BE LIKE THAT GUY?!" Luna shouted, and stared at Ti with fire in her eyes.

Quiet.

"Sorry, Ti, It's just that somebody has to care." Luna's voice dropped to just over a whisper.

"I know, Luna. I think of the news sometimes too, but Terra is home."

Luna had a solemn look on her face, but the innermost amber made the green in her eyes glow.

"Don't worry sister," Ti said as cheerfully as he could without making things worse. "Soon enough you will get your wish, and until then we have a tennis game to finish."

"Ha, good luck. I just got that new racket with a hundred point win on its name," Luna replied as a small grin spread on her lips.

"I better save my energy for after work."

"Yeah, like they have you doing anything important at the Base."

"Hey, you might be amazed at some of the stuff that I've seen."

"Yeah, when are they going to let you touch some of that stuff?"

"Soon enough Luna, and some of that stuff I might not want to mess with anyway." It was Titan's turn to look at Luna with burning eyes.

"Well then here's to soon enough, but too soon for me to beat you at tennis, though."

They both smiled at each other, and got ready for the day. They both worked the swing shift from noon to six, with occasional overtime and around six days a week. They readily preferred this schedule, but what they did not prefer was the switch to graveyard shift every five months. This was also review week and they knew it would unfortunately happen soon. They moved slowly for this reason and both reluctantly walked into the small elevator which always moved too fast. Each sky scraper also housed a central bank of many elevators to speed human transportation. Each pod would only fit four passengers and would rarely stop to pick up anyone else. This was also because of the speed at which they moved; a speed which many people would complain about. The pod stopped at one of the multistoried car ports, and they exited very slowly. Looking around they were astonished to see that only one commuter car was sitting in the middle of the spacious area.

"Looks like one of us might be late. I can't believe there's only one car left this early in the afternoon." Ti said as he looked around once more.

"I have an important meeting at the metro administration building," Luna replied.

"Why do I even ask?"

"You didn't ask because you already knew," Luna said playfully as she punched him in his trunk like arm. "You might as well drive, you're faster." And she opened the driver side door for him.

"Thank you sis."

After Luna got in, the car automatically read each of their security id's on their wrist computer (they referred to this small computer as a WR.A.C 'rack'- wrist access computer). As the two passenger car folded down, moved the seats into their settings and the familiar female voice greeted them. *Good morning Titan - Alluana. Titan now has two hundred fifty points on security license.*

"Wow… how did that happen big guy?" Luna asked.

"Hmm, maybe some things do happen sooner than later," Ti replied with a curious look on his face. "Looks like we will make it on time- activate security clearance 506."

"*Yes sir.*" And the cars pearly white melted into a deep black with silver shields on the front and sides, the small fog lights normally light blue turned to orange and yellow.
'*You now have fifty points on security license.*'

They were silent as Ti squealed the tires and wound his way through the exit ramps and onto the street. The city was moving as Ti started turning and weaving his way through traffic, while the small commuter cars were all two car lengths apart going the same top speed. This was possible with distance and speed sensors in roads and cars which reduced points and money if 'over the limit'.   It wasn't perfect but it did reduce crashes and eliminate stop and go traffic.

Luna felt calm in her city and looking around she thought of how great it seemed; this giant marvelous city.   How all of the people were outside doing something in such great harmony.   Even the noise had a flow, a constant moving eddy of people.   She saw an old lady slip in one of the many commuter bridges spanning buildings and a few people immediately helped her.   This almost brought tears to Luna's eyes, but she thought of Titan sitting next to her and she tried to blink back the wetness. Everybody was willing to help one another to keep people flowing.   These winding roads were incredibly busy but all seemed to move as one.   Her mind was floating and she could still hear the laughter emanating from the Acropolis walls.   It made her smile to think of how everybody can coexist with such differences.   All she ever heard of the world was not present to her at the moment.   She heard of the outside world as being 'square'.   It had crossed her mind on occasion as they passed the uniquely shaped buildings while looping in and out of tunnels and overpasses on their crazy winding rollercoaster ride.   They never stopped until they got to her destination.

"Great, now I'm early," Luna said, with a big smile on her face. "Hopefully we can still play that tennis game later."

"Yeah, I will ask to keep my new ride. Love you, sis."

"Drive safe. I will see you soon."

She watched him peel away and speed through traffic, and thought to herself that maybe things shouldn't change so fast.

The Administration

The Administration building didn't look too big on the outside. Just seventy levels up it was a dwarfed building, but it also went nearly fifty levels underground. This building also housed all of the foreign embassies in apartment form. Many countries didn't like this and refused to have an embassy unless they had their own building. The administration of New Terra simply said that they didn't have the land to accommodate such demands. Luna felt lucky to have her office on the top of the building in the third floor.

When Luna turned the last corner to get to her office, she was greeted by the newly appointed International president, Marie Corba.

"Hello Miss…"

"Please, just call me Luna."

"Well, you sure break through formalities quickly, Miss Luna."

"Yes, it always makes speaking to people that much easier."

"That's exactly what we need right now, Luna. May we go into your office?"

"Of course and what may I call you, Miss President?"

"Marie... now we may as well get right to the point of things."

Marie explained that New Terra had been taking heat by the international communities. Luna's application for International Ambassador for New Terra was being expedited. For the next six months she was to be trained in international law and policies. And within a year she was to be traveling around the world to explain how New Terra works and why it works so well. This is exactly what Luna had wanted to do for the last couple of years, but now it seemed strange to her that she felt so indifferent about her new job.

Security

Halfway to the Base after dropping off his sister, Ti got a strange feeling that something very big was happening or was going to happen soon. The feeling left just as quickly when he saw his friend Marcus speeding onto his road from the onramp with his own transformed commuter patrol car. When Marcus noticed his friend he grinned, they locked eyes and nodded at the same time. The race was on as they slammed down the throttle to weave around cars and corners. Halfway through the Compound, Marcus turned on his patrol intercom to try and mess with Ti.

"Hey buddy, good thing your sister isn't with you. She would weigh you down and ask me on a date after I win." Marcus's voice rolled and he seemed so calm and distant.

"Ha, I'm surprised you still don't have that shiner she gave you."

"That just proved that she wants me."

"Yeah and she sure helped you see just half of that 300 pound tourist lady from the United States of fat."

"Hey that was a nice girl who knew how to have lots of fun."

"I almost feel sorry for that cute chubby girl, hanging out with scum like you."

"You hang out with scum like me, and eventually Luna will realize that scum is the fat of the land."

Laughter … "Not likely, the only way you're going to win is if you keep making me laugh."

Marcus sounded more serious, "I still talk to her. She has applied for citizenship, and just so you know she's down to 175 pounds, and I like curvy."

"Well, now you've made me feel bad. I guess if I…"

"Yeah, yeah, everyone knows you and your sister complete each other. Kind of weird if you ask me."

"You know what we have been through, and it's not like that."

"Hey, someday I will thank your sister. She helped me open one eye at least, and she helped me keep my mouth shut… You win. I've been called down to the reservoirs on the south side. Got to take the next exit, it's faster."

"We will finish this at the racetrack. Then I will buy you and your lady a drink someday."

"Sounds good bro, see you soon."

Marcus left the main road into the Base, and turned down towards the reservoirs. Titan thought of his sister's devotion to the world, and how she just wanted to help everybody and know everything. This made Ti feel bad for the first time in awhile. He should know better than what he said about that sweet, chubby lady from the States. And he should not stereotype the rest of the world so easily, even when you don't hear anything good about the world. Anyway, all the people of New Terra were originally people of the World. He knew this was a fresh country but wondered where and how his life would have taken him from a different past. He thought of his sister and how she would probably punch him a good one if she knew what he was thinking so he expunged it from his mind.

When he parked and got out of the car next to Hangar 2 he was immediately approached by his commanding officer. There wasn't much said. Everything seemed strict and Ti was told he was to head down to Hangar 30. This was a hangar that Ti had never been to before, and now he was really curious if something was wrong. He wondered 'Should I be worried about my sister?' 'Was he going to be able to play that game of tennis?' He decided to try and keep his thoughts to himself, and even harder, put them away until he knew what was going on. He tried to not think of anything, but he couldn't. 'Graveyard shift wouldn't be that bad, right?' 'It gets old, but you almost start to like it, right?' He suddenly realized that he was crossing the Base at a crawl because the hangar roads are all square and that he has to stop at each one to look either way. He finally realized that the rest of the world is made square and blocky. He tried to remember the last place he had gone in the Americas and he started to laugh. He knew it wasn't too funny but he couldn't stop by the time he reached his destination. He popped out of his car with a huge grin on his face only to be frowned upon by many guards. They looked too serious but so much so that it was not funny, like they had just gotten out of a funeral for a dear friend. Titans grin soon turned around and he was cleared by the door guards. He entered Hangar 30 to see Mark Colburn, the newly appointed President of Finance surrounded by techies and mechanics. Some of them he recognized but never really got to know. Huge tables surrounded the group which was stacked full of blueprints. The president had a notepad out and was writing down whatever the techies were talking about. After a minute Titan felt shocked as he finally noticed what they were talking about. A massive curved dome was halfway peeking out of the ground. The sphere was so big that it seemed like the earth was giving birth to a new moon. Ti then saw a massive section of curved metal being dropped onto its unfinished side. He wondered what in the world this massive ball was built for. As he was staring into the unfinished section of the sphere, trying to make out what kind of machine this was, a female voice called his name.

"Titan,"

He turned around and replied, "Yes, ma'am?"

"I was told you are like your name implies."

"I guess so, ma'am."

"I am your new commanding officer. My name is Samantha, but you can call me Sam. You will report to me from now on, so sync your Wrac with mine."

"Yes ma'am."

"That's Sam not ma'am from now on, okay Titan?"

"Yes Sam, but please call me Ti nobody calls me Titan." They studied each other as they synced their wrist computers. The form fitting Wrac was small but could project holographic buttons and displays, and it would automatically put each other on a list of contacts.

Samantha had a cute look to her, and Ti could not tell how old she was, but assumed she was older because of rank and attitude. She was toned and strong which was easy to see even through her baggy jumpsuit. He wondered what ethnicity she was, because she was very dark of skin and hair, in contrast to her bold, light eyes.
She on the other hand noted how chiseled his features were, like a marathon runner, but not malnourished in build or stature.
Their Wrac's beeped and she broke the fixed gaze as

she turned and motioned for him to follow.

"The pressure from the international communities has gotten to the tipping point," Sam explained as they walked out of Hanger 30. "Normally this is an everyday event, but we now know that there are spies that slipped through the cracks in new citizenship."

"Spies? From where?" Ti asked as curiosity peaked, and he fell in tune with her every move as he could not help but look at her while walking beside and one pace behind.

"From several countries, we think .. six people, all from one family of refugees."

"They must have been trained well. Where are they being detained?"

"We have no idea where they are, they vanished off the grid."

"Impossible!"

"What's worse is that four people have been found dead."

"Murder in Terra? You can't be serious."

"Does it look like I'm joking?"

"No, sorry ma'am, I mean Sam."

"The Domestic President Liza Delany has given me unrestricted access in order to stop these spies as soon as possible." Sam stopped, turned, winked at Ti and asked, "Have you been to Hangar 25?"

This was when the shock started to fade and he realized he was standing at the door to Hangar 25.

"One of the few hangars I haven't been in," Ti replied as he looked at the giant building in front of him. He silently questioned his own answer.

"We do have technologies that powerful countries want, but when it comes down to the dirty, well sometimes you just have to do things the old fashioned way."

"What... like killing people?!" Ti asked trying not to sound too alarmed.

As she waved her Wrac in front of the door and stepped inside Sam replied, "Maybe but only if you're put in that situation, Ti. We would like to detain these spies, but if you must kill somebody then….are you the man for the job?"

Ti's mind was clear. He could not focus on the unknown, and he stepped through the door.

"Good, you're not the only one. We are assembling

a team that you will work with."

"Not my sister!"

Sam smiled for the first time. "No big guy. Your sister is not apart of this. She is safe."

Ti wondered where Luna was, and if she knew about this. Then thought of the tennis game that would have to wait, and when he realized actual murderers were on the loose, he focused and his resolve was clear.

They were walking past rows of weaponry and body armor, as Samantha briefed Ti on the people they were looking for. Three computer engineers turned to hackers, two doctors which probably augmented their facial features to get around surveillance, and one mechanic, who was most likely some kind of Special Forces. Sam then explained that he was just one of five thousand patrol officers looking for these spies. With another twenty thousand computer techies, bio scientists, surveillance officers, ex security civilians, and domestic affairs officers who knew about this.

He was to patrol the outskirts of Metropolis which didn't have much surveillance, and the corridors surrounding the manufacturing Compound. He was given a lightning gun with five million volts in focused bullet form, body armor, his own modified patrol car, and a ten inch bowie knife.

A knife is what was used to kill four innocent civilians, and for what he thought? 'So another country could try to be like New Terra? Why didn't they just ask? Why would they put a glorified stereotype on New Terra and then want to hurt us?' Titan asked himself these questions as he got into his patrol car and sped away.

Titan's Vacation

Titan wasn't worried; he really had no care at all. He did not expect to be the one to encounter these spies or to even get a call for that matter. He always tended to be the last one to show up to the fight, and the last to receive training at Security, which sometimes would be the best way, as they had more time to give the training. This was not necessarily a vacation as they spent the time waiting for the call doing martial arts or shooting practice or other tactics in detainment before and after patrol. The 'vacation' was the change he felt away from his sister as he focused on new places and people. Luna had been prodding at relationships and games which had annoyed Titan lately. The constant irritation of a game that he thought he was good at was gone, as he was now a defender of his country, a defender for their freedoms.

Titan felt free as he was laughing, playing with the nice girl at the burger express. She was blushing as he and his partner had a hard time getting past her bosom with their eyes. She giggled as his partner,

Reece, a young-chiseled jock, fresh from boot camp and full of vigor he was asking her out.

"You should go to a club with us this weekend, c'mon it will be fun." Reese asked pleadingly as he leaned over Titan.

"I'm not even old enough," the girl replied with a disappointed look.

"Soon enough though right? Maybe I will see you there eventually the club is called Viper's Den."

"Okay." The girl was trying to hide her smile as Titan pushed Reece back while saying their good byes.

"You're terrible, but you're also almost her age so maybe you will see her soon." Titan said this as he thought of the women that he should try to go after.

"You almost sound jealous, you must have lots of ladies, and don't you always go to these clubs?" Reece responded in a very mature voice compared to just a moment ago.

"I go to the clubs because of patrol, and they hook me up later, off-duty. And yeah, I have women in my life." Titan said this and thought of his sister. Coincidentally he felt the vibration alert of his Wrac. Titan knew it was Luna and he ignored it as they drove along their route and back towards the base.

They took this time to get to know one another, and to eat. Security had just trained a 21 year old physical masterpiece who was Reece. He had also been orphaned as a child, but his parents were never known to him. He had lived in America first, before an unknown cousin had filed the citizenship forms to take on a permanent vacation from his large foster house. They didn't even notice his absence he proclaimed, and he was proud to be a citizen.

Titan finished his own brief past, mostly with the relation to Reece's as they made their last looping street from the end of their shift.

"Damn," Reece quickly burst, as the night shift was closing, "I'd like to have some action... fuck those spies."

"You really think they'd let us newbie's take all the glory? Nah, we will be the last to know." Titan remarked wearily.

"It does feel like training," Reese lowered his brows and squeezed them together in a look of attempted anger towards the situation.

"Hey, you're training with the best you may as well be fighting spies." Titan looked at Reece and thought of a brother at first, and someone who creates ruckus at the bar second. Or maybe it was all just determined patriotism. In any case it was late and Titan was turning off the road and would technically be off-duty in seconds.

Red lights flashed on the emergency alert. For a moment Titan thought the auto scan had malfunctioned close to the gate. The dispatcher came on almost shouting.

"All S.S.F MEMBERS THIS IS CODE 712 RED-ALERT, LOCATION IS SOUTH 8041, BUILDING C, SECTION 23. THIS IS SOUTH CENTRAL COMPOUND AREA. I repeat. ALL…"

Reece looked at Titan in fear of his decision to enter the base, to run from danger. Titan glared at Reece with his own anger for a long second and turned the wheel toward the street. Screaming wheels ignited an adrenaline fire in both men as eyes were squinted in determined anxiety. Chills of unknown death and pain lingered into Titans veins as they quickly entered the large Compound. Close enough to Security to be a martyr's mission; soon the place would be crawling with Special Forces. And too late, agents were already entering the building with the big 'C' overhead. This relaxed Titan as they closed in and exited their own vehicle; their car made it four with six agents already inside. Titan rounded the car before he realized that Reece had already entered the building at a sprinters pace.

"Wait…Damn." Titan ran inside to be witness to another closing door from a long hall. "Idiot, you're going to get yourself killed, this is not how we train." Titan rambled to himself before he realized the map on the wall. 'Section twenty-four, this place is huge.' The interactive wall map was an excellent display of each large manufacturing area. Section 1 to 23, but section 24, 25, and 26 stood out, it was the largest and contained the actual machinery of this plant. Most of the other rooms were accounting and engineering offices.

A step forward... and a decisive step back made Titan question the actual location of the fugitives.

Section 23, engineering, was also the place that had thermal cameras as his security clearance automatically showed this to him on its display. The places that had no thermal cameras were manufacturing areas. The machines were hot and would not allow any precise location of any abnormal body heat. This is where the spies would be if they are still here. Titan knew and felt this as he walked into the large manufacturing facility. Many smaller areas were still full of automated machinery with design rooms surrounding these spaces. The sections were moving slowly so Titan began to jog down the large main corridor. It seemed eerily quiet, and Titan could not help but worry about that young Reece, who really wasn't much younger than Titan. 'Too fast and head strong, it would get him killed.' Titan shook his head, slowed, and braced to open the main doors to section 20. He reached out to push the doors open, only to find that they already were, as he nearly stumbled forward into three handcuffed people. They looked pitiful and beaten but it was laughter that struck at Titan.

"Almost got here in time to catch one Titan... too late."

Three pairs of security agents each held one captive as they laughed and pushed Titan aside. Reece was the last to enter the hall with a disgusted look on his face as he glared at Titan. Reece almost wanted to follow the other agents to maybe gain some glory, but stood and stared at Titan as he waited. Titan waited for privacy before whispering angrily to Reece.

"Three more spies idiot. If you run like a fool then you might run into the killer, do you want to die?" Titan crept back into the door while Reece's eyes got big before he followed with his hasty response.

"They aren't in section 23, surveillance has combed for heat signatures."

"Those were doctors couldn't you tell. They are most likely in sections twenty four through twenty six, probably trying to hack the engineering grid in manufacturing, this way."

Reece followed Titan through an engineering section 20. Each section was most likely owned by individual contractors. Each a driving force for good parts and better pay. The main room to section 20 had a large round table with floating 3D blueprints. Heavy doors closed and silenced the automated machinery behind them. The still quiet made each foot fall sound like knocking wood as they crept around the large room and peered into each of the smaller offices. Section 20 through 23 was telling only of vacancy as Reece was the first to enter the corridor to section 24.

"I see movement, Titan."

"Wait."

Reece did not wait and they both hastily entered the giant manufacturing facility to unveil many automated motions. Far away noises greeted them in a mashed jumble of robotic actuators, humming motors, screeching machine tools. They crept along the side of the wall before choosing one of the many aisles between large machines. It didn't seem long before Reece's sharp eyes picked out more movement, which this time happened to be human. Running he pulled out his stun-gun and fired on one of the shadows standing in the middle of hot working machines. Titan followed as Reece again ran after the flight of one of the enemy. 'Stupid,' Titan proclaimed as he ran but slowed to notice the convulsing twitching body with a stun bullet that landed square in the spy's chest. Titan also noticed three small computers on a table hardwired to a larger computer server. Titan decided to move and he disconnected the small computers in swift fury. Turning to follow and find Reece, Titan was almost past the row of machines and it was strange through noise and vibration, but he felt movement from behind. One of the spies had stayed behind, hidden behind a large tooling cabinet. Titan crept slowly noticing the silhouette of a muscular frame. Muscles bulged under black clothes as the computers were already connected with keys being pounded furiously. It was the first time he had ever unlatched the holster to the stun gun outside of training. A snap of the holster made the spies ears twitch but no other movement. Titan tried to be fast and civil.

"Put your hands on your head and kneel down or I will shoot."

The man kept typing.

"Do it now." Titan reaffirmed in a stern voice and Titan pulled the gun out and took aim.

The man stopped and raised his hands, turned and spoke in a raspy uncomfortable voice.

"You're too late; your beloved country will be left in the dark."

Titan did not think anybody could hack Terras' mainframe. He did not want to take the chance and aimed and shot two of the three small computers. The spy was faster then expected and closed the gap with an arching roundhouse kick aimed perfectly at the stun gun. A third shot fired but too late as the gun was booted from Titan's hand. The turning spy brought a barrage of punches to Titans face.

Defending as well as he could from the unexpected onslaught. Then a moment of peace as Titan let his guard down, and too soon. The hard right hand of a built-up motion created a shadow of darkness behind Titans eyes.

Titan woke to find the spy standing at the one remaining computer, cursing under heavy breath. Titan felt a mashed face as he slowly rose pulling the concealed knife out of his boot. *'Now it's on'.* Titan rose tall and relaxed. *'You will not harm my people.'* Titan thought this and thought of his sister. Rage gleamed under slit focused eyes. He calmed his mind, thought of how calm his sister was under any situation, how fast she was- this spy did not compare. Titan closed in and spoke.

"Get the fuck down on your knees, or I will fuck you up."

The Spy turned, noticed the blade and produced his own in a quick slight of hand.

"I wasn't going to kill you, but now you've pissed me off."

Titan's rage subsided and he realized that it might not have been the best idea. *'Be Calm'.*

The lunge was fast, but not as fast as Luna as Titan backed away and circled feeling the man's speed. The second lunge brought three quick slashing arches all narrowly dodged by Titans grace. The spy looked annoyed and changed speed. Backing Titan around a space between hot machines the spy crouched slashing upward then down as Titan blocked the knife with his Wrac. Relentless stabs and slashes, paring, finally cutting deep through the Wrac into Titan's wrist. Titan had had enough. The next upward slash was caught by Titans bleeding limb. Pulled upward so the blade faced outward and Titan let the man feel his strength. Titan's own knife pierced the man's stomach, starting a slow rise toward the man's heart. The spy spoke, grabbed Titans hilt, and nearly leaned forward into the blade.

"Feel my death, your people will feel it soon."

"Why do you hate us?"

"You all live like kings while the rest beg for scraps."

"I do not want that."

Titan started to pull the knife from the man but was met with a push and heavy grunt. Titan could feel his blade against hard muscle, pulsing, another heavy grunt and the man pushed the steel into his own heart. Blood poured onto Titan's hands, arms, and stomach. They parted and Titan dropped the bloody blade. The man finally kneeled while bursts of blood poured onto the floor. Titan also kneeled and placed a hand on the man's shoulder.

"I do not want this." Titan wanted to call for medical help.

The spy looked hard at Titan in his last breaths, eyes glared big as wet glass. The smell of life's blood rose thick as strong incense in Titans senses. A white mask like face, augmented and unnatural looked sad in the shadows of light, but the man smiled.

"It's a good death; I am a hero to my people. Nobody else has breached your walls."

"Why do you hate us? We mean no harm."

"The greedy hide behind strong walls. What are you hiding?"

"We hide nothing."

"You should all… live like … so… many.   Others."

Titan shook the man but he collapsed in the puddle of blood creeping large and reflective on the concrete floor. Dark glassy eyes stared back at Titan, and Titan could not look away.

***

Two weeks passed since Titan had dropped off Luna, and Luna was waiting for her brother to return home. She had gotten calls from her brother every other night, but he couldn't tell her what was going on, which angered her- intensely. She began to constantly ask her superiors to tell her what was going on, and she finally got the news out of Marie, the International-affairs President. The last call received was Ti calling to return home in half an hour. Twenty five minutes had passed. She was angry.

Luna heard the apartment door beep and slide open, and she ran to her brother, tears already welling up in her eyes. Titan stepped through the door and was immediately squeezed by his sister. Ti gave a firm hug back, and told her not to cry, which made the tears fall that much faster.

"Did they get those bastards?" Luna asked sniffling.

"I got one. The rest were detained."

Luna stepped back and then realized his left hand was bandaged half way down his forearm. "What do you mean... you...?"

"Yeah but only in self defense." Ti couldn't look at his sister's swollen eyes, so he looked at the floor.

"I was worried but I'm so glad that you're okay." Luna started weeping.

"Everything is fine now, and I love you too."

They held each other for what must have felt like an hour.

**

Three months later, Luna's studies were nearly complete. Her office had been moved into their apartment so that she could be with her brother while he was on medical leave. The two cuts from the Special Forces spy had been deep but could have severed a hand if it hadn't been for his Wrac. Mentally he was left with scars as well but not nearly as bad as his hand, for when he thought of fighting that trained killer, he just had to imagine defending his sister. Death was apart of life, and in this life, nobody was going to hurt his beautiful sister. As he thought of these things a small grin crept across his face, and a small tear dissolved back into his eye. He then made his sister a pot of coffee.

"How's it going, sis?" Ti asked, as he walked out of the kitchen and put down her coffee on her desk in the living room.

"Not bad, just finishing my studies on ancient and colonial Mexico." Luna paused and took a sip of her coffee.
"I can't figure out who was worse, the Aztecs or the

Spanish conquistadors?"

Ti knew when she was asking a serious question. "Well, one was more sophisticated, but they were all still human with religious fanatics on either side."

"Yeah, hopefully they are more like Terra now. They got to be, right?"

"Maybe, people are weird, people in the news always fight… want to see what's on the news?"

"Sure Ti, I need a break anyway."

They both sat down on the couch in front of a wall in their apartment. The wall at the moment looked like a 3D jungle moving with ferns and trees, which parted in the middle into a valley vista. Very deep and detailed like a Mona Lisa without the Lisa obscuring the view. Luna pressed a couple buttons in air from her Wrac and the image faded into the breaking news cast.

"Widespread panic, as people across the world will wonder if they will survive the winter without power." Ti and Luna looked at each other, and Luna turned up the volume.
"If you're just tuning in, there has just been a catastrophe of epic proportions as eight out of the fifty nuclear super reactors have just gone through melt down. These two year old reactors where supposed to be the next generation of nuclear fission.

The Unistar power company hailed these reactors as cleaner, more efficient and safer than any other reactor on the planet. Critics said they were built too hastily with too many cut corners, which seems as if, at the moment, the critics were right."

"What's that?" the reporter put her hand to her ear piece. "I'm sorry, citizens of New Terra, the news keeps getting worse. Four more reactors have experienced slight meltdown while they were being shutdown. So far reactors in the U.S, Mexico, China, India, and the United European Union, have all experienced slight or total meltdown. With us now is the head professor at the New Terra Institute of Technology. I wish we could have met on a different occasion, Professor Lazara."

"Yes that would have been nice; unfortunately I was one of those critics who is surprised that this didn't happen sooner. Greed Becky, greed makes the world a more terrible place as they not only cut corners but they tried to push more power which the cooling jackets, though technologically advanced, could not withstand."

"How does this affect New Terra?"

"Not at all, as we use totally clean gravity based proprietary generators."

"And by proprietary you mean..?"

"That New Terra is the future of energy, just like

how New Texas, I mean... just Texas was a big energy supplier of oil at the beginning of the twentieth century."

"So we are going to be selling proprietary information to the rest of the world?"

"Depending on what the countries want to spend. We already have machines that we can sell which will make it very hard to reverse engineer."

"Why would we want to sell this technology when people need it now?"

"The rest of the world has shown that it doesn't care, therefore New Terra might have to also create the technology to clean this mess up. That's why our founders made this place. They were tired of all the politics that just wastes time, money, and eventually-people's lives. The founders wanted to create one last Eden, one that can be a symbol of life and prosperity for the rest of the world. And if the world wants to accept us, then they should be willing to pay for their mistakes."

"The rest of the world does not have freedoms like we do, though."

"And that's where the world's governments need to start. If people start to play our game then things might be different, but they want to make us play with their archaic, greedy, forceful nature. So that's how things may play out, but let's hope for the best."

"You heard him here first- Professor Lazara, on Terra's first news network channel three on the tree. I'm Becky McDaniels. We will be back with more coverage in three minutes."

"I can't watch anymore," Luna said, as she turned off the screen and looked at Ti.
"I thought this world was done destroying itself, Ti."

"It's stupid, people who don't care is what's destroying this world."

"Yeah, and it's been happening for too long, and I used to think that at one time everything would be fine," Luna said looking and sounding sullen.
"That's because you live in New Terra, sis. It's unfortunate that you will leave to see the world in such a bad time," Ti said as he rubbed her back.

"I'm not sure if I do want to see it anymore, or at least, not parts of it."

"It can't all be bad... Let's go get some lunch, okay?"

"Okay, but I want sushi..."

That meant they were going to have some sake, which is what Ti wanted. Sushi wasn't his favorite food. He actually didn't like fish at all, but didn't mind the eel or some of the dumplings. He never told that to Luna though, and she barely ever drank alcohol, only on occasion and usually just at the sushi bar.

New Texas

If an idea had a place in history it would be or not be. Those willing to strive for a greater good may be called ambitious at the least. Many thousands of selfish things people may do if they truly want to live and prosper. Energy was one of those things that New Terra would capitalize on for the survival of all mankind. Thousands of ideas were put to test to find the greatest renewable energy resources. The New Terra administration was steadfast in its right to capitalize on these technologies, especially in reaction to the worlds 'blind eye affair' with large corporations. As long as you have money nobody asks questions, which is what the founding billionaires of New Terra had relied upon.

One founder was from a family in Texas. They were oil and solar billionaires. At the right time Alexander Jones was also surrounded by genius, two of which were his own children. The network of billionaires in the corporation that eventually formed New Terra had many genius people actually discussing the future of the world and its fate.
The reasons at that time were more of corporate

greed, which in the long run actually paid off. They were discussing what they were to do after the earth's vast oil supplies actually dry up. How solar, or wind, or nuclear could be used to power all electric housing and auto. In his old age, Alexander only cared about what would be best for his children. He believed that nuclear could be as bad as oil to clean up. This was pushed through by cost over waste, over clean up economics, and the other corporations agreed. They thought they may as well take the road that is cost effective over time, and secure a place at the top.

That was in 2050 when a giant corporation poured money into a, 'research and stress proof account fund.' Three years later the catalyst for New Terra would also be a savior for the world.

Autism meets Savant-ism.

Alexander's children were gifted in many ways, but his autistic child, Adam, would be the last person to be called savior at the right time and place. He was never drugged like many autistic children. His father thought of this as disturbing his natural flow. His son sparked life into his father when he was born, and being able to afford treatments of mind and not chemicals could have been key to unlocking potential. Little Alisha Jones was his second child who was also very gifted an all aspects and raised with the same compassion.

They were raised with the same fortitude and love as the twins were in the unwritten future. Little Adam Jones found to be autistic at an early age did not make his father feel bad at all. It had in fact brought laughter and spirit back to the old man, who thought he would die bearing no child (he was only 45 but life takes its toll). When his children were of a sociable and perceptive age is when the 2050 research fund was approved by the giant corporation all owned by nearly 200 billionaires. This meant his children could have access to all of the world's technologies no questions asked. And in 2053 when his child was just 24, Adam Jones had become the world's most genius astronomer. His autism was focused into -savant-ism- and to him looking at math and stars was like looking at a beautiful giant Picasso. His mind could read distance, and trigonometry, and sub divide those properties with things like gravity and momentum. When he focused with his powerful telescopes he could almost read stars and planets like they were maps on a bearing with coordinates. He loved looking into the sky, and putting math into what he was looking at was easy. Everyone around him was awestruck at the genius of things that would never cross ones mind. He and his sister were being taught by the best, and most could learn something from Adam. It took him just three years using high powered telescopes to spot the distant rock that could potentially hit earth.

Two weeks after first spotting the rock he said something to his sister. She had just gotten to the observatory to pick him up…

"Adam, are you almost ready to go?"

"Nnnnnnoo ready yet sis."

"Well you should be. All you do is look up there… what's new?"

Adam took his eye off the telescope and made his look where he couldn't quite form the words because he could not quite come down from the astronomical plane in his mind.

"EARTH'S GONA DIE!" Adam said this so loud and quick it was like a shout to most people. He was happy he said it right, and he smiled with bright eyes and looked back into the lens.

Alicia knew his brother was never wrong when it came to math, it was his language. Her hands felt numb and tingly. She tried to feel for a wall to brace her but she only felt air. She was dizzy and could feel the black hole of sleep creeping up her spine. She wanted to kneel, and she could feel her legs give out from under her…

She could feel herself moving…

"SSiissyyyyy…'''

She was being shaken when she woke up.

"Nnnnnnnn… SISSYYY!!"

"I'M AWAKE, STOP SHAKING ME PLEASE!" Alicia sat up. "Stop, stop, its ok Adam. I'm fine."

"Wa ppen sis."

"I can't think of what you think of, or else I go to sleep." Alisha stood up slowly and rubbed her head. "Lets sit down, Adam, ok?"

They went to the observatory's break room. Alisha made some juice, and they sat down.

"How is earth going to die, Adam?" She cringed as she asked.

"Big big big fast rock… comet."

She could feel the blood drain out of her face. "And when is this going to happen?"

"Fa fa fifty-eight years…" he looked at the clock, "and and and two thousand eight hundred eighty seven hours twenty three minu…"

"OK, ok, ok, please don't tell me anymore Adam. I, um, can't tell time like you can, remember?" Alisha put her head in her hands. "And you're absolutely sure, Adam?"

"Yeah, three teles in three hundred hours."

"Why didn't you tell me that was the reason I'm driving you everywhere?" Alisha gazed at her brother. "I thought you had a school project, or something."

"Naaa ad to make sure… teles tuned."

"You know that's not good news…" and then she thought about the dinosaurs. "Maybe… at least we know."

"Yeah sis."

It didn't seem to bother Adam, which Alisha thought was kind of weird. She didn't want to ask what he felt though. She knew that he had emotions, but his terribly brilliant mind kept many things null and void. Then she realized she would be in her eighties by then, and probably dead anyway. After all she was taking a flight in a week, which could also turn into a rock, falling to the ground. And then she realized she was being selfish and thought maybe people should know, might as well do something about things like this. She gave her brother a hug and a kiss on the head and decided to call her father.

Alexander Jones was skeptical, his first thought was to call NASA. Then he realized something very powerful in this situation. He knew people were their own worst enemy, maybe NASA knew and didn't care. The Space agencies had grown lax over the last few decades. Most of their profit was just suborbital flights for millionaire's kids. Alexander realized these agencies were not even big enough to handle something of this nature. And- he realized that scaring the world at this time would just make things worse. He told his daughter to stay quiet, and to get the data that Adam had. He then called a few professors in various fields that he knew, including those few astronomers who worked with his son. He told them all that they would meet at the observatory to validate an astronomical discovery made by his son. He told them to be quiet until it was validated, and that they were to have favors and a dinner party afterwards. He knew everybody would make it because he was rich and his party favors were not cheap.

The object was tiny on the enhanced imagery screen. It only looked slightly lighter than the space around it. They were cross referencing data from another observatory which helped calm down Adam in a strange way. There was a moment when the computer had put a cleared path through our solar system. But Adam was quick to explain the slight gravitational pull while it traveled across the edge of the Milky Way in its last ten years. The astronomers agreed the system was not up to date. It hadn't taken long to get an update and make a couple of tweaks and the group sat still as they watched the advanced timeline. Adam tried to explain it took him a week to realize where this thing is going and what might be in its way. This was his excuse to be somewhat normal but everyone knew he was a walking computer. This chunk of rock was heavy supernova traveling at an incredible pace. The rock looked elusive, like it did not want to be seen. That is one thing that peaked Adam's interest, but the rock's speed and power was still not a problem for Adam to comprehend. Alexander was indifferent when the computers arcing lines finally reached our solar system. He knew his son was right and was proud of him for it. He had already put security programs on all of the systems to keep this quiet. Sometimes the less people know the better. The group didn't want Adam to be right. It didn't seem like the lines would cross but the computer finished and the lines stopped. Alexander was the first to congratulate his son. Soon everyone was clapping slowly with alarmed or grave looking faces. Alexander explained to everyone that now is not the time to tell anyone, and worldwide knowledge would only stress the world. It could create even more depression in the fragile economic and social kind of way. He said he would talk with some people and figure out a financial strategy to pay for a 'cure'. In a week they would be notified what was next, and until then they were to celebrate Adam's discovery.

Two weeks later Alexander was sitting in front of a group of billionaires with a net worth in the trillions. His short documentary explaining how gifted his son was, and why they should keep his discovery quiet impressed them all. They seemed more interested in the ability to calculate vast sums of whatever you would like with no computer.

"Why did you never tell anybody about your son?"

"He was never pertinent to what was being said."

"And you're serious when you want to try to buy out part of Mexico?"

"Not only would we save the world, but if we are smart, and we are, then we can make lots of money."

"Unless people have knowledge of this, we should probably start 'losing money' unless we want to be buried in political taxing, am I right, Mr. Jones?"

"Correct, and in a few years when the world's economy is in dire recession... we will save it with a new state in North America. This is no joke and we will have to approach this cautiously, and with certain people. I for one will pose as an arrogant old fool who thinks that his grandfather knew of some vast oil deposit in the sand."

"This might be crazy enough to work. I can back you as a manufacturer who needs a place to sell at high prices. The Mexican and U.S governments will eventually see financial gain and will probably be expecting it to fail anyways. What kind of politics are you looking into?"

"I am not even going to worry about that right now. What I do know is that we must try to cater to all who help out our newest Texas. And unlike my home in Texas this place will be small."

"You're right. We must all stay ignorant to complete a common goal. We will all be dead in 60 years or less anyway. And if there is no world then… you all know what I mean."

"How this plays out will truly be a human test. It is fortunate to see how people can master this concept without it being a problem."

Alexander could see the closure in everyone's eyes. He knew they would live their lives to its peak. His billionaire friends were ready to pressure the governments that they hated more than anything. He too would be happy if he could put his money into something better for his kids. He wondered if his kids would witness this event, and hoped that they would. He hoped that the power players in the group who asked so many questions would stay vigilant. He could not do this alone. Money was only relevant if you had enough, only useful when put to work. He hoped they would be willing to secede into a new country, a place with no political gain because the only gain was in this world, into a fresh new Texas.

Chaos- 2110

The year 2110 was nearly coming to a close and so was Luna's trip around the world. She had been able to schedule many tourist locations around her ambassadorial duties. She felt like her job had not been taken seriously by many governments. Part of her job annoyed her as well, and she was content with seeing many things. She was told to tell the governments that they would soon see a documentary on the value of New Terra. New Terra was also going to send some research vessels into space. Most countries did not care because they knew that New Terra was a research and development capital of the world. This made Luna feel as if her trip was more like a vacation on her way back to Mexico in the Americas.

Mexico was her last stop for the next month and she was happy to be close to home. She wanted to see her brother even though they spoke with each other nearly every day. She wondered what he was doing and hoped that trouble was staying away. And in a month she would also know the details which had annoyed her so.

Titan had joined securities Special Forces unit and had not told his sister so that she would not worry. He knew that something big was going on in the last six months, and he had a hard time comprehending some of the things he had seen. He was also told that he would know soon enough and that he shouldn't worry about what nobody knew. His ranking had gone up drastically and he pushed for it, but this did not make him feel better. He also thought he was smart until he had seen some crazy machines and even crazier blueprints which made no sense. His worry was always more directed towards his sister but she had guards and she seemed happy when he talked to her.

He had also seen his commanding officer, Sam, on occasion when she wasn't busy. He had never really had a woman, other than his sister, around his apartment until now. He almost felt like he was cheating on his sister in a way, but he knew in his mind that it was a ridiculous thought. Sam was always on the move as if she did not sleep and it made him wonder about what she knew, and what he didn't. It did make him realize that Terra was the place to be in this day- there was no place better.

Two weeks later Titan was assigned guard duty south around hangar 40, close to the reservoirs, and the border. He had the feeling like he was defending something when he looked out in the desert. He also knew he was looking at where his sister was coming from and he wanted to see her badly. One week later he was assigned to do part time training in hangar 14. This week was intense and Ti was strapped to various flight testing apparatus which he mastered almost naturally. Half of it was like playing a videogame and half was mathematical basics.

Busy, was the only reason he never asked what was

going on. Everyone was so serious, and he wondered if they knew or if they were just being silent like him. He knew his sister would have pulled the info out of somebody, she always knew the right people. Luna would be home soon enough but not in time to hear this news together.

The day that everybody was to stay home and listen to the news was the first morning that Samantha had been to Titan's apartment.

Titan heard a curious knock at the door.

He turned on the surveillance imagery and saw a stunning lady in a formal skirt looking almost like a librarian. She looked up at the camera and he crossed the room to open the door.

They tried to talk at the same time. They were both making odd gestures while in vague speech.

After a moment Ti asked, "You're going to tell me what's going on?" and motioned for her to enter.

Sam entered in gauche silence. She turned and abruptly tried to say something, but stopped and asked. "Can we sit down?"

"What's going on?" Titan asked again.

"I should have told you sooner, but people are different when they know."

"Everybody is different lately. What could be so bad?"

"It's the worst possible thing Titan- complete global

annihilation, from an extremely destructive asteroid."

"What? That's it? Everybody has seen asteroid destruction movies... are you serious?"

Samantha started to laugh hysterically, "You haahaaahahahaha you can't teashshahaahaaa... you can't not care. Because it's true, but we..." her eyes were big and they seemed different, like she had been up all night and couldn't sleep, "we might survive."

Titan was not looking funny which made her try not to laugh that much more. "You really expect me to believe this, huh? Are you joking?" He tried to sound stern and serious.
Her laughter was non stop and Titan was getting annoyed. She could see his distain and she started crying while laughing. He did not think he would ever see such a strong woman break down like she just had. Ti sat down next to her and thought of his sister and how he should handle this.

"You... you don't want to know what I know anyway... nooo," and Samantha started crying on Ti's shoulder.

Ti realized she was telling the truth and he was determined to be strong. "So, we are all going to die?"

"No, not if our machine works, but it's a real killer... a bad thing."

"So tell me about our machine, and this stupid rock... your bad thing."

"Soon it will be close to our solar system and we will all see it much more clearly."

The Rock

This anomaly was an ultra-dense heavy chunk of destroyed star. It was a piece of two or more collapsing stars called supernovae. This piece of rock was not able to complete the transformation into a black hole and its repulsion is what created a wave of exploding debris. It had even traveled through lights boundaries on a strange dimension at the beginning of its time. It was small but very heavy, and it wanted to travel fast. It consisted of something that was more like anti-matter and anti-light, like a glass that you could look through. It had seen the most terrible of cataclysmic events and it had lots, and lots, of energy from it.
This fact made it elusive to most eyes that scanned the sky. It didn't look like anything because it was an anomaly. Its speed was astronomical to someone who could not comprehend its power. Fortunately most interstellar bodies of earth or heavy plasma would not even notice as it shot by on its stellar

highway.   The fact that somebody found and calculated its path was also an astronomical event.

Adam had seen the shot through the barrel long before the sniper pulled the trigger.   To Adam it seemed like earth was just in its way, wrong time and place... too bad.   He realized that this rock would not only destroy Earth, this rock would completely devour Earth and leave nothing behind. Its gravity would turn Earth into a pebble in comparison to its own mass, and if anything, give it more power. This posed a problem for the engineers and astronomers who were to take on this task.

Alexander Jones was the man to put down 100 billion to create New Terra. (He still left his children a few billion.) He cast his snowball to show that the other billionaires should not worry about any amount of money anymore.   There were greater risks in this world that were not man-made.   Risks that showed no mercy for greed or poverty, things that can be taken for granted. They followed him because they knew he was right. This world is what made them all rich, and what good would it all be as nothing? He died shortly after his first team was assembled for the prevention of earth's annihilation.   He died happy to know that his children were making strides through earth's politics for mankind's benefits. He died knowing that his money was being used for a greater good, even if it meant that they couldn't fix this problem. He wanted a better world even if the worst was to happen. It was not something that he thought was subterfuge; it was something more intrepid, and valiant. He was happy to see the first ten years of this new country, and how it flourished in the dry and arid landscape.

By now his daughter and many others had taken the reigns, which were heavy to carry. Most had no idea what they were dealing with. They had been busy building a utopia on top of a giant secret. When Adam disclosed his information to these scientists they had no idea what was in store. Most had never heard of anomalies like this much less knew any math about cosmic physics. Adam would try to explain that they could not destroy this rock. The best chance was to either slow it down or throw it off course. A nuke would not even be noticed by this thing. This would have to be much bigger than anything on this world or else it would be like hitting a bear with a flyswatter.

Adam explained that this is not a rock but more like a chunk of star which didn't know what to do with itself. He tried to explain that even gravity, one of the most constant of energies had no place here. If anything it repelled other gravities because it was more like a wake in space. It was moving too fast and they had to stop its progress before it would hit our solar system. This depressed many very brilliant scientists. They could not comprehend this kind of power, much less try to create something that would even budge this thing. Many gave up in a binge of booze and depression.

These things were subdued as there was always hope to be found in numbers and in many people's minds. It came ten years later, right before the twins were born. They had new technologies for energy which made it easy to focus their efforts to release it. Heavy uranium and other nuclear material was going to be a catalyst to make a most deadly and unstable material. This material was going to be shot out on a stream of light energy to make its purpose useful. This light was going to be high frequency gamma radiation, and the material was going to be converted matter, anti electrons, or antimatter.

This material was the only thing that would even scratch this body of power. The rock was already closer to antimatter than anything we would perceive as normal matter. An ostensive material was going greet a most unforgiving taker of life.

A chaotic event was going to happen in one form or another.

Eye in the sky

Luna could not believe what she was seeing. The embassy's assembly room was packed around the big screen and everybody was quiet. They were watching what the rest of the world was as well, a documentary on the formation and purpose of New Terra. She had seen the true founders who tried to explain what they were up against, and crazy footage of places in Terra that she had never seen or even heard about. This made her emotions peak and she felt betrayed by her own country. At the same time she did not want to be at the embassy anymore. Luna wanted to be home with her brother and their friends. Her commanding officer told her to follow and get the debriefing for the questions that everyone has. He said he was sorry, but it was kept quiet for a good reason, and that tomorrow it will be all over. She was to watch a series of technical videos about the rock and New Terra's machine. She got half way through the first video while the autistic man, Adam, cut into the footage trying to explain this event in a crazy mathematical language. This made her feel insignificant and small. Tears were rolling down her cheeks as she thought of the little cherub angels that adorned the Abbey and Acropolis. She thought that Adam kind of looked like this angel, and that he was like a little angel of immense insight and unstable power. He was the only real being that could come close to the comprehension of this devastation. She made herself watch, and when the video became more technical, she wished she could know more about the true founders of her home.

Three hours later Luna was nervously trying to answer questions in front of a Mexican panel of government officials and scientists. It made things worse when she tried to explain that she also knew nothing until now. How could she not know what was going on in her own country? They demanded answers. She told them what everybody knew and that they would see this tomorrow morning at 5:40 live coverage for the survival of humanity. She was very pleased when she left to do media coverage. The scientist who was sitting down at the panel to relieve her looked pale and in the last place he wanted to be. The media room was not any better. Luna tried to be professional and spoke of how she did not need to know because her life was good, and that all this was for life and nothing else. Coverage lasted into the night- for those who now knew were probably not going to sleep. Luna finally got out of work and the media glare when she got a call from her brother.

"Ti! What the hell?"

"I just found out too, sis. Can't be mad at me."

"Yeah, well, so this might not even work?"

"We will know in, what, around a year."

"Damn it, Ti, you know more than me. What is it?"

"I've seen crazy machinery but everyone's busy. They have giant spheres which are supposed to be the saviors in the sky. I am supposed to be on one of these but it's a different security version."

"What, this isn't war, right? What the hell?"

"We are trying to deflect a giant asteroid, Luna. It's going to be a big bang, I'm told. It's going to be loud and really big."

"And if this doesn't work than we are going to do what, Ti?"

"I don't know… something hopefully. I don't have any answers, Luna. I'm sorry."

There was a pause for a moment.

"I wish I was home, Ti."

"I know, Luna. I've missed you too, but you had a good time, right?"

"Yeah… the world is a big place, Ti. I hope it lasts forever."

"Hey, the spheres are going to be launching, I'm glad you're close- and in three, four hours we will know, right?"

"Yeah..." Luna did not sound happy.

"Hey don't worry, I'm only afraid of your tennis game."

"Ti… I already won that game."

"I didn't even want to know sis… you're much better than me."

"Not when you try Ti, but thanks, I love you.

"Have to go Luna, love you."

Titan was shocked to see these giant globes lift off the ground. He felt pressure on and around him and he felt small, unlike his name, or anything that he had been told. They all wore special ground suits for the electrical field which he thought was amazing. He was being struck by lightning and it did nothing to him, him or the thousand other people standing around him. The spheres were creating thunderstorms around them. He realized this was powering them in a way. They lifted one at a time out of a massive storage bunker under the ground. Rain started pouring as the last two emerged, deep shiny-black in color, unlike the rest which were shiny silver. One of the black spheres was mostly flat looking on the bottom of it. The other dark giant had a flat top as it crept into the air. They moved as one connected string of electrified globes climbing slowly into the atmosphere. This connected string of giants stopped over hangar 50 which was the largest hangar at the edge of the base. He had always wondered what was in that giant building. Out of the forty one hangars in the base, this was the most secure. At that moment the loud speaker came on, *'Attention all medical and S.S.F with clearance 8 report to hangar 30, I repeat, clearance 8 medical and S.S.F report to hangar 30 in ten.'* Titan had just enough time to see the entire building of hangar 50 roll back to reveal part of a massive dome. He wondered how such a giant could hover like the rest as he made his way towards hangar 30.

Three commanding officers were talking as they stood silent looking at the giant ball. They were apart of a newly formed rescue team. Each medic was assigned an S.S.F member. Each team had their own mini-mag jet. They were told that this sphere was of last minute design. This ball would not be near the blast but would get one of the best views in the 'park'. This was a transport and medic ship it was not militarized. It did not look finished, and Ti could see what looked like circular vents on the sides and bottom that he could not see while they were airborne. The man who started shouting was a Commanding Officer of Security. Titan remembered him from boot camp.

"WE are going to save the world ladies and gentleman. This is being overseer by the brightest minds in the world. WE will have a front seat to witness this world's survival, and WE WILL SUCCEED. Each of those eight IRL Spheres has just two crew members- a pilot and copilot. They have just one jet dock. The Black ones are Diffusion ion reactors with a high frequency thermal wave emitter. The techies just call em' Black Dire for short. Those two have nobody, they are remote controlled, and your questions are as good as mine. The big one is my favorite and is called Forward Amplified Positron Burst Transport or F.A.T. The big also has nobody. I've heard some techies call this operation the BIG DEAL and some call it TAFFY. I call it operation FUBAR. WE are not the ones to bow down. Whatever it is, it's going to wish it had never been born!"

"S.S.F MEMBERS FOLLOW ME, DON'T WORRY, AND REMEMBER YOUR TRAINING!"

They entered through the small port on the bottom and climbed into a long small chamber. The seat he was placed in felt like the flight chair in training, and he realized they must be facing outward. The medics entered after their own C.O. and sat behind him in seats he did not know were there. They all had goggles and flight masks on, so Ti had no idea who was sitting behind him. He realized they probably both thought the same thing.

How did this happen so fast, and how did I get in this position. A screen turned on and everyone could see the hangar wall.

'SYSTEMS TEST…OK'

5

Ti couldn't believe how fast things were happening.

4

3

He realized this was a last minute preparation, and he didn't feel prepared.

2

Where was my sister?

1

Lights flickered and he could hear inner workings taking place. Gears meshed slowly at first and electrical started snapping. Ti relaxed when things started moving faster, smoother. He was tense again when things started to sound like explosions. He wondered why he was here. The last explosion made the sphere move and the sounds started to become harmonious. One thing was finally moving another, and he felt buoyancy as they rose into the air. The screen reappeared with a view of sky and a human face to the left of everybody.

"Greetings, everyone, sorry we put this team together on such short notice. We had priorities and didn't know if we could build another so soon. We have a couple of jets but we of course have limited airspace- until today. All the techs have agreed that this blast may be bigger than everyone anticipates. This means we need to make sure the pilots in those balls are safe afterwards. Each squad will be assigned one of the eight. There are sixteen of you. You have been trained on how to fly them. It should be easy. Your HUD will tell you where to go, it just depends on who needs the first one out. Right now they are taking position in our outer-atmosphere, it may take some time but we will definitely get a good view."

There was movement behind the face, and the man moved out of the screen. Ti suspected he was some high ranking tech. He knew there were thousands of people involved, and he realized he was one of the thousands who had his place even if he didn't know the details. The screen started focusing in on a giant storm cloud miles away. It then filtered the cloud and focused on the lightning. The thermal image was amazingly crisp and it looked like a little solar system of static and thunder. The screen then filtered the storm completely and the giant spheres looked strange floating so high in the sky. He felt movement and the screen jumped as it was refocused. He felt strange when the craft moved but realized that it must move fast. He imagined how there was thunder and lightning around them too, and how it will be fun to pilot in a brutal storm. They stopped just as fast. The globes on the screen began to show geometric lines intersecting one another. There was one giant ball below the two balls in the center. This massive ball looked like the star in the system. He wondered - three times as big, four? The eight were close and slightly above the same level, surrounding the giant. A line shot out of this giant and through the top of the black spheres and out of the screen. Another screen popped up scrolling numbers with what looked like two layered graphs. The small simulation sped up and the lines had eventually crossed. A yellow light appeared on the graph.

A couple minutes later a voice told them to move to another position and that their reference coordinates were critical for secondary information. They rose higher into the atmosphere, and Ti started to feel more weightlessness. He wondered if this was going to work, and for the first time he felt real doubt. The voice came back a couple minutes later and said they are making last minute tweaks. The same graph-plotting sub display came back on and did a couple fast simulations which all showed green. The voice was loud, "HOT IN TEN MINUTES, BLAST IN

FIFTEEN."

Titan wondered what time it was. He realized he had talked to his sister just a couple hours ago, and things started going fast. He hoped she was doing well, and he was glad she wasn't involved in this. His seat felt like jelly and now he realized why they were so comfortable. He wondered how this was going to feel. He thought of shooting guns at the Acropolis and how he liked it, but not as much as shooting arrows.

"HOT IN FIVE."

The screen with the giant balls moved and two more screens appeared. One screen showed the graph screen with numbers scrolling which only a computer could register. The other screen was live unfiltered footage, as massive black cloud with constant lightning flashed in the shadows of cloud cover. He thought he felt a pulse from his heart matching the distant thunder. Then he felt kind of stupid. It was this ship making that pulse. He probably couldn't hear that other storm.

"HOT IN ONE MINUTE."

The graph screen slowed down and it looked as if certain numbers were being tweaked at a tiny level. Ti wondered how much math was being crunched at this time. How many computers and people constantly reviewing until…

"HOT IN... TEN, NINE, EIGHT, SEVEN, SIX, FIVE, FOUR, THREE, TWO, ONE..."

Titan's mind was clear and focused. A beam of light rose out of the cloud. A flash of light made the beam brighter and Ti felt a shockwave like he had never felt before. He looked at the filtered screen and all of the giant spheres were eclipsed with light from thermal laser beams cooking the two black giants. The black ball on top was reflecting the eight beams toward the sky. The black ball on the bottom was being cooked by a very intense light from the massive sphere beneath it.

Screens appeared which showed different light spectrums and magnetic propulsion systems. The lasers were getting hot, and what was the black orb was now a shiny beacon of brilliant luminescence. This light seemed unnaturally bright. One of the image filters was a close view of the two hot orbs melting together as giant lasers merged. It took what felt like a full minute... the two orbs fusing together and a reaction took place. The lasers went super-hot, ultra-bright and Ti knew this was the blast....

This was the moment that could save earth... The techies still wondered if this would work. Most did not even comprehend the destructive forces which they had created. The bomb would be an ultra reactive antimatter burst on a stream of light-radiation which would chisel a hole in a black anomaly.

The explosion did not look like anything because it was hidden in a flash of brilliant illumination. It faded into a shockwave of a growing Sun which was fast devouring anything in its growing bubble. He was going to be disintegrated. He could feel its approach, and Titan's heart jumped. The shockwave seemingly powered by a blooming sun and he felt crushed as the ship went dark and was slapped with a tremendous deafening force. It felt like being booted, tossed, and carried on a wake of power. Red emergency lights were flashing and propulsion power started its pulsing rhythm. People around him cursing and moaning as their momentum felt terribly downhill. It seemed like too long before the machine's thrusters made its familiar buoyancy feeling. There was a slight sigh as people regained composure. Screens started turning on one at a time, and Titan noticed how bright it seemed outside. The flash had taken the clouds away completely, and a ball which resembled a small sun had leapt out of the mortar jar which began its course into space. A terrible, large, red text came on screen flashing- #8 EMERGENCY. The screen was focused on a massive ball which didn't look bad for being on the ground. Then he noticed the charred remains of the two mortar shells. They were glowing and seemed crystalline. They collapsed on top of each other and had pulled the giant sphere to the ground, melting onto its side in the sandy desert. The image moved to the other grounded ball- so much smaller looking. Ti felt movement and the two people sitting next to him were gone. He noticed tiny trucks on the screen speeding towards the massive downed giant. They looked ridiculously small. The screen changed to yellow and focused on a floating battered ball around a swirling cloud- #4 ALERT. His seat moved and he blinked and was sitting in a tiny jet. He felt the familiarity of the simulation and he pressed some system checks. His HUD showed his coordinates and he pushed the thruster forward. They were slammed backwards and were shot outwards screaming through the air. The rain didn't look like anything, and he realized they were filtered by their goggles or the craft's windshield. He could see the ground and he righted his craft, centered on the coordinates, and pushed the thruster lever all the way forward. His comm. link crackled on, *"Now it works? Damn, ok, please don't speed up so fast, ok?"*

"Sorry I didn't know how it was going to feel either. Not as bad as that blast though."

"Yeah I'm starting to wonder if I'm the right person for this."

Ti realized it was a woman speaking, so he tried to calm her. "Don't worry, they picked you for a reason, and we need good people."

"I'm only useful if we don't crash, ok."

Ti imagined he was talking to his sister. "No

problem. A baby could fly this thing, and I'm a pro."

"Ok, champ, as long as I survive I'll be happy. We must be close?"

"Yeah, we're close." Ti eased back on the throttle and lightning crashed around them as they slowed down.

The big ball looked like it had been beaten with a giant bat. It looked dark from being blistered and scorched. A wave pattern with stress cracks had formed around the laser
port on the top of the machine. Ti circled the spinning globe and backed into the jet dock.
They opened the hatch as soon as the jet powered down and the green dock light flashed. The medic threw Ti a bag, she told him to go to the opposite side of the jet to open the pilot's cabin.

'Turn the door lock counterclockwise 360 degrees. Pull and slide to your left.'
Watery goop helped the door push open, and Ti slid the door out of the way. A woman was laying on a jelly surface with a mask full of blood. She didn't look alive but he pulled her mask off and started to do c.p.r. She coughed blood when he pressed on her chest and she spoke, *"I hope you are one who I chose. Take my comp keys and change the world... Terra can not die."*

"What do you mean?" Ti asked as he realized she was probably in her eighties.

She convulsed as she opened her eyes and her last word was *"Titan"*. She stopped shaking and had a peaceful look to her.

He got chills and stepped back. He couldn't look away from the woman who knew him, but they had never met. A portrait formed in his mind, a woman with bloody beads coming out of her eyes, and ears, and nose. Her mouth was completely red, and he imagined her as a young woman. He imagined that she could have been his grandmother, whom he had never known. She looked serene and happy, even through the immense pain which she must have endured. A pain which he thought could have been avoided. Why was she the pilot to this ship! He realized someone was speaking to him.

It was his medic teammate, *"Do you hear me? Hey speed racer what are you doing?"*

"Yeah, sorry I was trying to revive her. What do you need?" Ti moved quickly as he said this and opened her wrac and took the two comp keys and put them in his own.

"This guy needs medical that I can't give him. I need help moving the med bed onto the jet."

"Ok, I'm coming. Didn't know we could do that, might have been nice to know."

"Sorry, things happen fast around here. Let's do the same and get this guy help."

Ti crawled across the jet's hatch as she pressed some buttons on the jelly table. The man melted into the gel and the box closed. Ti noticed that the mask was not nearly as bloody as the woman's, but he looked beaten and comatose.

"He's alive, barely. How's the other pilot?"

"Not good, I don't think she..." Ti looked away, and felt defeated.

"Its ok, we should still take her."

They pulled out the box and pushed it under the wing of the jet. The latch closed and they gave it a tug to make sure it was secure. Ti felt strange when they did the same to the woman, like they were burying someone that they didn't know. Curiosity filled Ti's mind and he almost wanted to say something to his medic, but knew that he shouldn't. They crawled back into the cockpit, and Ti prepped the jet for takeoff.

"Please don't accelerate too quickly, ok?"

"I won't, I know how she flies now." A plot was set on their HUD for the nearest ground support crew. Ti took off slowly and the craft felt like it wouldn't accelerate too quickly anyway. After they docked with the ground crew, they were relieved of their medbeds and a new course was set to help more people.

Alluana had a very different vantage of the super-weapon. She returned to her embassy apartment where it was too quiet. She turned on the screen immediately and found a decent music channel. She didn't want to be alone but she also didn't want to know about this- any of it. She realized her life had been better off not knowing of her own imminent demise. She loved life and hoped that it would last forever. She fell asleep with tears on her eyes and hope on her mind.

Luna wakes to feel the earth shake. It looked like daylight outside, and she ran to the window. Luck would put her looking north into a new sun. This sun grew bright for its first breath, too bright to look at, and it was gone. Two minutes later she was still looking at it careening out of the heavens. Her eyes were strained as she tried to blink away the halo of light. It was getting smaller fast, and she felt proud knowing that her country had done something that nobody else had ever thought possible.

Luna looked down for a brief second to see what looked like a veil of mist. She felt some of the shockwave before it fully hit. Things started to shake and a rushing gust of sound and energy smacked the area and kept moving. Her initial instinct was to duck down, but she was on her back before she knew what hit her. When she rose she was in the corner of her room with glass and dust on her face. She looked up to the ceiling of the room which had a massive split in the wallboard. Everything looked like it had been moved or thrown around. Luna dusted herself off and sat on her bed while trying to think clearly.

She called her commanding officer and said she wanted to go home.

Siege - 2111

The world had turned to grief and anger. People finally felt the tipping point in survival, as the world was already struggling with a great depression. It was going to be another year before the world would know if New Terra's weapon would even work against such a deadly beast. The blast would collide with the silent rock just outside of earth's solar system. The sun's heliosphere would help deflect such a burst of unforgiving energy, and nothing could be predicted except for something astronomical.

People were angry over the silence and the unknown. Silence was the best thing that happened for the last sixty years. While people bickered over nothingness, calm endured. This calm could not last forever, and a stressed out civilization finally would know of a possible worldwide destruction. It was a catalyst for lawlessness and anarchy. A deadline in life put many on the brink of insanity. Many blamed New Terra for such an outbreak in anarchy. A scapegoat was to be had, and most wanted to cull the shepard of prosperity.

Mexico was the first to demand answers because their land was used as the blast site. They were moving military camps near Terra's southern border. Tanks and jeeps with guns poised for takeover. New Terra's politicians were dumbfounded that nobody cared about anything worthwhile. In a time of utter human reliance everybody was being self-centered and terribly venal. It was clear now that government was not being governed by any democratic means. Corrupt influence had long since destroyed any true independence or freedom. The only place that resembled a true democracy was the place that wanted to survive and thrive.

Titan's Gift

Two days after the shock that the world still felt, Titan entered Alisha's lavish apartment underground. He had not spoken to his sister, but had been told that they were on their way home. When he asked who spoke with her last, he had gotten no answer. The news was in the dark for a few days. Many communication relays had been fried from the electromagnetic burst of Terra's weapon and so Titan did not feel too worried.

The inner door read Titan's wrac access keys and welcomed him inside.

'Please have a seat at the screen after you are comfortable.'

The environmental system recognized his preferences and adjusted for him. He grabbed a glass of water as it seemed convenient and sat down in front of the large wall-screen. Video started with a strange music mix and images of desert and water. He sat back and watched.

"I cannot apologize for the late information. The world has too much already, and what good is one more piece of knowledge for the wrong people." The bloody woman he watched die appeared on the screen. She looked very happy and beautiful. Chills ran down Titan's back and he stood up.

"By now you must know what Terra has done." She looked down, "I would hope that my memory is felt by you soon. Hopefully we can save the world and hopefully all is well for you and humanity. This is our home, and Terra was created not just to save humanity, but to save our home." The screen switched to a picture of Earth majestically floating in space, and her voice still cut Titan to his core.

"I hope you like what Terra has become. It had been over thirty years in the making, and it is amazing what people can accomplish." Titan realized that this was made years ago, and she must have been in her fifties.

"New Terra has grown like my father said it would. This is fast becoming the best country in the world. Your legacy will not be in vain."

The video showed images of Terra being built in fast motion. Her voice followed the images. "Unfortunately New Terra will not be built for erroneous prosperity. It seems like our great minds are creating more than a purpose for survival. We are creating the greatest nation in the world, too quickly for this... archaic age. It must be done though, for survival's sake. I love this world, even though it is way too corrupt to be 'good'. And the corrupt nature does not like prosperity. Wars have all recently been waged over money and power. This means that Terra must lay low before her time to shine. Powers and governments have been moving against our way of life, and I don't know if that will ever stop in this volatile time."

Images of massive hangars were shown, and construction of giant machinery was taking place. Images passed before Titan's eyes and information had been downloading into Titan's wrac.

"I can't expect anything, but hope for the best in this world's prosperity. I also believe that every citizen should have the chance to uphold freedom from terror and injustice. That is why I wanted to be a pilot... so that I can help save the world." A beautiful smile and Alisha had aged considerably. A picture from his past fell into the screen and Alisha was standing next to his parents each holding a small child. Tears fell from wide eyes as he already felt like she was his grandmother. She was talking about the glory of New Terra's people as Ti saw the security screen flash red with two people approaching. A small light above a corridor was white and Titan feeling wet cheeks headed for it and didn't know why. As he headed down the corridor a few people entered as he exited into a small room. In the room was a small tunnel with a shuttle, and his wrac beeped green and Ti entered.

The door closed, and the small gray craft started its descent in a snap. Titan was weightless and falling... faster for a time that seemed too long... then a slow brake with a quick turn and the ride was over.

He crawled out and didn't know why he was here or what he should do. He did know that his sister was still not home and that he wanted to know what was going on. Moving down a small corridor he heard voices and slowly crept into a small control room with a few people looking into... something. He slowly crept forward and the room seemed to get bigger until he was staring at the base half of a giant globe. This ball looked similar to the ones used in the explosion but was smaller- still menacing in size and seeming very dull and indestructible. He walked until he could see all of it, and then realized what the people were looking at- an old looking screen with old scan locks which suddenly beeped green as Titan crossed the scan line. The people jumped, about five of them.

"What the Fuck?" one of them said.

Titan didn't care as he looked at the screen which read -

-Security updating, entry accepted.

-Defense control activated.

"Sorry if I scared you," Titan said.

"Whatever." The voice was from the middle most person who sounded like his copilot from a couple days ago.

"You didn't scare me, just made me jump that's all."

The guy who was next to him slapped him on the arm. "I'll be damned, if it isn't Titan."

Ti looked at his old friend, Bruce, from a mechanic's class, and he smiled.

"Hey I didn't recognize any of you either... until now," Titan remarked as Wrac syncs updated to post these new peers, "so what is this, the holy grail of globes, or what?"

The lady who was jumpy responded. "It's one of the first prototypes for the Lsats emitter program, or in laymen's terms- big fucking laser. The first like this was supposed to be weapons grade. Let's take a look."

She led herself, two women, and what looked now like four men through the catwalks and into the landing pad.

"This is ultra-classified, so if nobody is here that means those who knew were probably the pilots from the other globes."

"So, we are all the new crew, eh?"

"Looks that way, the other pilots are in the hospital..."

"Hey... another corridor, want to take a look?" Titan's curiosity made others follow and they entered another dark room and a small scan light turned on. Their wracs beeped and they read a message- CLASSFIED Return to LS Infinity program. Their glances back through the room felt ruined in darkness. They could not even tell what size it was as they walked back.

"I used to think Terra was supposed to be a small country," Titan proclaimed.

"A small country with big things living inside, and speaking of which, let's do this." The small familiar woman led the way into the side of the sphere.

Titan was last to enter the globe and took a place in seats that felt familiar in shape, harder in substance. The woman who was the medic now sat in front of Titan. She turned and looked at him.

"Are we good v man?"

"V Man?"

"Velocity… ah never-mind… are we cool Titan, I'm Mindy?" she turned as red peaked though her cheeks.

"It's a pleasure to see you Mindy, and will take it slow this time…" Titan said, and thought 'Hopefully' as an afterthought.

"We all need answers, who are all you people?"

The ship's screen turned on at the right moment and welcomed them all with a smooth deep voice.

'Good day- Officers of the Free Republic and Society of New Terra. The citizens of this world must need you now… My name is LS Infinity one. My purpose is- weapons grade. My system is upgrading…

I am to welcome Mindy, Vera, Wendy, Bruce, Nicolas, Luis, and Titan. You have all been promoted to Commander of operations, with security clearance 10.'

Their comps flashed with a new grade and downloaded security clearances.

'The Infinity project creates two things. A high intensity constant laser. And a directed magnetic resonance burst. Either of these things will boost human survival by 85%.' Schematics finally popped up on the screen and the ball was shown with four

laser ports on the bottom and four on top, with similar systems as the big orbs, but on a smaller scale.

Four of eight IRL lasers from Human Survival Initiative... Updating... This craft is weapon destroyer. This craft will leave New Terra's boundaries with your permission...'

They all pressed the button on their Wrac.

'Infinity one is operational...'

'The southern Mexican army has been surrounding New Terra. Updating... Tanks positioned toward Security... They will also cut flow to the reservoirs with one of their battleships.'

The craft lit up as its motors started grinding. 3D screens dropped down in front of the pilots with more holographic information. Titan had a hard time reading the flowing tutorial but he started to get optimistic and read faster as he felt calmed down. The craft seemed easy enough to pilot as anyone of the members could take control to defend or protect. It was a 360 degree floating weapon The humming sound started its melodic drift, things smoothed out, and weightlessness was produced. Titan was weightless, the craft moved, he knew he would soon see daylight. Or at least through all of the different filters on the many screens which he was trying to focus on. Targeting... systems... diagnostics, systems from the other four ships started pouring in on a small map on a globe, and the view that was approaching his gaze. Tanks and soldiers looking upward at billowing funnel clouds twisting and flashing through the dense mist.

The central screen caught attention which showed the desert and the Mexican army. They had organized a large military siege wall. Tanks and trucks were slowly setting formations. An army of soldiers were busy setting up fortifications or digging bunkers.

Terra's other ships had just been briefed and had gotten orders.

'Comply with Infinity One and resume transmissions.'

The small craft, comparable to the rest had lifted and started to maintain with the other craft's magnetic cushions that the massive floating globes produced. Infinity one then rested on a cushion of magnets easily, for when the volatile explosion occurred the giant globes had produced extreme quantities of energy, pressure, and electromagnetic 'shields'. Four giant globes where surrounding him after their final ascent into the clouds, while a large Mexican army was facing them in a five mile wide crescent shape.

The Globes where floating high in the sky above the compound and base that had been evacuated. No doubt the people of New Terra were quite frightened. Massive thunder storms had ensued for days and everybody was on 'security' alert.

A few hours later the formations of the Mexican army had stopped. It had seemed like a scare tactic, a display of power. Titan tried to keep his mind still, wondered how he had gotten into this position. His sister was still sitting beside him... but she wasn't.

He had always known that he would see her again, though... he didn't think it would be in the company of barbarians, wanting to take back the new country.

Mexico had gotten rich and prospered because of Terra's prosperity...next to a seemingly small country in the middle of nowhere which created many of the medicines and expensive goods of the world. Terra had long been dormant to Mexico through the decades. Mexico still maintained lawlessness with corrupt power struggles, but one man's bane was another man's economy, and Mexico still prospered next to Terra.

Now the small forgotten border produced a large army. It was menacing enough, like a three pronged trident to Titan and he knew his sister was not well just then.

His gut wrenched and he fell forward into his screen.

"Hey, you ok?" Mindy was looking at him.

"Yeah I'm good, just got a stomach cramp."

He took the controls and focused around the army trying to look for something out of the ordinary. It didn't take long to spot the activity at the front, a General was giving commands and soldiers were setting up short wave radios and podiums.

Others were focusing at the front now and international COM links and hotspots were being diagnosed from Terra's security communications. A general took the stage and within a moment started speaking towards New Terra and the contained funneling clouds.

"This has always been Mexico's land and long have we prospered. Now you build a doomsday device and we do not want it on our land anymore. SURRENDER!

The general looked restless, nervous…"We are now closing your reservoirs and we have the first victims of this war… they will DIE if you do not comply! BRING THEM!"

The monitors focused, and they brought out five people from a large truck.

Alluana appeared in the middle, and Titan focused onto her. She looked almost dumbfounded, in a strange place, and surrounded by an army. She had her public-relations clothes on, and it seemed like it was just yesterday, for a second, and the cold truth crept through Titan's veins. He noticed the General put his hand on his gun and Titan whispered, "Luna."

The gun slid out of the general's holster. Titan was

calm and started speaking,

"Everyone, that is my sister and she will not die. I will not let him pull the trigger."

The pilots of the four massive globes responded, "Copy that sir. Copy that sir. Copy. Copy that."

His Copilots responded. "Copy"

Titan had the General in his sights. The general's gun was pointed in the air with fingers rapping on the handle.
The general was standing in the middle of the group and he started to speak.
"Will it be... this one?" The general lowered his gun and tapped Luna on the back of her head.

Titan whispered, "No."

Titan tuned down the heat resonance to less than one percent- pulled the sticks trigger. A large hole appeared in the general's forehead and a heat flash seared his face in a brief instant.

The gun flashed. Luna rolled forward.

The other pilots locked onto the surrounding area and started to fire. The laser port at that side of the craft made small thumping noises.

The tanks surrounding the Base were incinerated.

This alerted the Mexican army but it was already too late.

The large Globes also locked onto the tanks, weaponry, and men. Four flashes of light burst out of each of the ports. The lasers were big. The four had a giant, two-foot, diameter beam which was now also tuned down to less than one percent. The Mexican army lasted less than thirty seconds.

Crippling explosions from melting dirt and metal, so fast in its destruction, most had no idea it was already over. The energy and heat charred massive swathes of land. Skeletons were made and destroyed in a wave of heat and radiation that was still cooking after just a brief instant. The bombs and ammunition added to the strange annihilation.

Titan looked around for his sister after he destroyed what he thought could be a threat in the vicinity. He could not find her.

Missiles launched from Mexican bunkers and war ships. Titan and his crew were staring at the missiles, quietly looking at death.
Red alert lights were flashing and auto-locks were focusing on the incoming missiles. The person sitting next to him was a woman named Vera. She laughed and asked, "Are they serious?"

"It is a joke. Most of the planet is fucking insane,"

Titan replied as he noticed her looking angry at the missiles.

Vera smiled and whispered, "Let's show them."

Luis, a pilot a couple seats down added, "Let's finish this."
Triggers were pulled. A hundred small thumping sounds and sky was alight with fire. The twelve massive lasers floating in the sky had more than enough points in any direction to destroy any projectile or threat. The second volley of missile was destroyed much closer to ground as a warning of superior firepower.

His copilot from the other day found Luna for Titan and put it on his screen. They were still running towards security as base transports were on route to rescue. Titan took the time to message her back with his thank you.

After the attack things were quiet as Mexico and the allies were in shock. Streaming data and diagnostics never ceased, and the machines produced a constant pulsing rhythm. He made sure his sister was within Terranian borders before he sunk into his chair and stared in the distance through all his screens. This made Titan feel drowsy. He tried to stay awake which made focus worse. It felt like days since he last slept and he wondered how long it had actually been. He had seen Samantha first then at least half a day to find Alisha's apartment which led so quickly to war. He realized this is how all soldiers must feel-strange quick changes in life and death. Time seemed to drift as scrolling numbers passed across his eyes. He focused on a set of falling numbers which made his eyes drop and his eyelids followed. Darkness set in only for a moment. Then he opened his eyes and was standing at a door. Its number was 118, and he automatically stepped through and started walking through a jungle. Birds chirped and leaves rustled. As he walked, a cool breeze hit his back. He found a trail and heard muffled sounds through the distance. He started to run towards them. The sounds got worse and sounded more like screaming. He left the forest and stopped in a desert on the top of a cliff. A woman was in the middle of an ocean of sand. She was sinking. Her face was bloody and she looked up at Titan. She screamed, "HELP ME."

He blinked and was falling off the cliff. The woman had disappeared but he was falling fast.

Sand looked like a vast painting and he felt a thump.

Retaliation

Alluna received first and some second degree burns, the laser was tuned to its lowest possible output to kill her saboteur. The shot of heat and energy still rang her mind… to close for comfort. Her bandages had been removed and skin was raw like a bad sunburn where bare skin had been scalded. Most of her hair was cropped, and all that happened did not put her in a good mood. She had stayed in a nice room in the bases subterranean complex. She had still not seen or heard from her brother, just a few days but she was growing restless.

Her wrac beeped, and Titan said, "Hello?"

"Finally, where are you?" Luna said, and abruptly asked, "why can't I contact you?"

"I've been piloting one of the IRL Spheres. I'm a clearance ten now."

Luna's eyes got wide and she said, "Oh."

"Where are you? I'm going home sis."

"I'm at the base. Get me."

"Give me your locale and I'll see you soon, okay?"

"Okay bye." Luna smiled huge, and tears welled up in her eyes.

You might have thought they were lovers. They didn't hug each other until they were home. Then Ti realized Luna was wincing from burns and he pulled away. She had a hard look on her face, and he was surprised that she hadn't cried. He wondered if he should tell her about piloting the craft, and the start of war to save her… maybe later.

"I want to know everything," Luna said.

Ti smiled and thought it had been awhile. He should know better. "Sure lets go to Alisha Jones's place, I have the keys."

"Should I ask how you got them?" Luna smiled, but then realized her brother's eyes were on the floor. He was terribly quiet while they were in their apartment. Luna realized Titan had been put in some strange situations. She did not even like to think of her immediate past. Things had changed and she was a survivor. She realized everything would be different. Soon everybody would be a survivor.

Titan told her his story as they made their way to Alisha's apartment. First how he was a pilot to help save Alisha herself after the blast, and then how he found her apartment as his copilots had as well. He tried to explain the weaponry as fast and as coherently as possible. He then admitted that he was the one who had first pulled the trigger which saved his sister. She was mostly quiet as he spoke and soon enough they were standing at the door. Titan pushed the button and the door scan read their keys. They entered and immediately stood before five big men... bigger in height or width than Titan. They did not know what to say but were told that they could enter.

The guards motioned for them to go down the hall

and into the room which Titan had used before. Now it had become a small command post, similar to what they both had seen in Bases central command. A man noticed them and smiled. He alerted everybody by yelling, "LISTEN UP... We all need to welcome Titan and Alluana. They have recently earned clearance eleven and will become commanders when the time is right... thank you all." Everybody went back to work and the man walked across the room and greeted them himself.

"Good to finally meet you. My mother has always had high hopes. Almost makes a kid jealous when she tries to know others. Anyways my name is Jonah Jones."

They both shook his hand and said, "Nice to meet you."

Jonah paused for a moment in thought, and then said, "Please, let's get comfortable and watch the gathering of the United Nations. We just made the world afraid of us, when Titan saved you Alluna... well, sometimes things are no joke. This anti-war summit may turn out to be an anti-Terra summit as well. Don't worry, though. We have many cards to play."

Jonah began showing them the different interfaces which made communication so smooth and easy. He pulled up the Infinity programs interface, and they were looking at five 3D globes floating high in the sky. Luna was entranced and stared into the machines which may have saved her more than once. Eventually she noticed the small shapes which must be people in a control deck. She pulled away before she got emotional and focused on reading statistics and news broadcasts.

A few hours passed quickly and all had locked eyes on the international debate. Two of the three presidents, Marie Corba and Mark Colburn, were being escorted and seated in front of a large international panel. They did not look frightened or anxious in any way. It was being held in a conference building at the ASU campus in Arizona.

The first person to speak was the American president Elizabeth Sanders. There had been women vice presidents, but Miss Sanders was the first woman president, and she took her job very seriously.

"You know we cannot tolerate warfare so close to American borders," Ms. Sanders said shrewdly.

Marie Corba was quick to respond, "And, like you, we cannot have Mexicans, or anybody, enslaving or terrorizing Terranian people."

"Americans believe that New Terra has been like a golden thorn in the foot which should eventually be removed," Ms. Sanders said just as sharply.

Mark touched Marie on the shoulder before she could speak and said, "American's downfall was never Terra's doing. It was always weak regulation. America had suffered in economy long before New Terra was ever even made, and, as I recall, Mexico has always been mostly third world."

This made Ms. Sanders noticeably angry, but others had started their inquiries before she could explode. A corrupt system had never left America's economic and political structure, and although New Terra had helped the US, its decline was still inevitable. Eventually Terra made people realize that anyplace in the world could be nice with a tolerant government. Many people saw the US decline into a third world before New Terra 'saved' its still terrible economy. It was a hard time for any country to prosper, but many were becoming 'content' in last place.

When the panel was concluded it hadn't seemed like anything had really been debated or resolved. Most had no response when Terra's presidents asked about the two Mexican battleships in the Baja peninsula where Terra maintained its water. The Mexican president had not shown up, they chose to have the canal's dammed and Terra's reservoirs were at 80%. When presidents from other countries asked about the floating weapons above New Terra, the response was the same and indifferent. They said they would fly until the sky was safe, which didn't get the greatest response from some. Other countries subtly asked New Terra for money or aid. It had lasted about five hours which time quickly dismissed. The presidents had a quick escort out, and the two siblings had a view of their transport back.

***

They should have been more alert, but it happened fast. An obelisk tower had been erected at the edge of a town in Alaska. It took the Russian and US military a couple of weeks to build and deploy one of their own top secret laser programs. They were smart enough to assume that no modern projectile weapon posed a threat at Terra's advanced weaponry. They had quietly built this weapon and a towering lift had put it in position at the crest of their vision. The weapon was bunkered down behind a mountainous landscape, in a cold hostile environment. Unassuming as it was, Terra's vision had been placed more in its immediate location, waiting for another Mexican attack.

The Mexican government had declared war but was like a snarling beast caught by bigger prey. Eventually they knew Terra would take back the canal, but they did not even know what the rest of the world was doing. South American countries were hard pressed to give any aid, much like the rest of the world. The United States and Russia were being quiet and had not told Mexico a thing. They wanted to be the power in the world. They did not like this new country's presence, and they did not care about Mexico.

The weapon took position. It looked like a 'T' lifting out of the tall 3,000 foot tower. It rose another 900 feet before it broke vision with the floating globes. The globes were casually buoyant in the sky, lifted by their ionized thrusters and a massive magnetic cushion. All being powered by a synthetically created thunderstorm, harmoniously stable in its massive power creation.

The weapon fired just as the Terranian presidents entered the border. A green line broke through sky. It was aimed in the center and would have gone through one of the big giants had it not been for the magnetic cushion which acted as a deflector. Its deflection made the beam weaker and caused it to go through the smaller craft's bottom half. Red emergency lights flashed immediately, and for a minute Titan thought he was in one of these ships. The small control room went up in red and everybody started screaming at one another. The ten second burst had done its damage and the smaller globe came crashing down on the four larger ones. Their evasive action was actually to catch the smaller globe before it crashed to earth. They then lowered their entire position and took to ground between the base and the reservoirs. The communication room seemed disorganized and crazy as people had been screaming for damage reports and casualties. Amazingly it had not harmed anyone when it passed below the control deck on the Infinity one. The pilots were coughing from smoke and burning metal, and the scene from the cameras were chaotic in the least. Luna gave Ti a hard look and than a worried one as she looked into the screens. Titan looked for Jonah and saw him talking to a different command center. Titan ran over to him as he was telling the Commanders to evacuate the area and get the troops ready for a ground assault from both the American and Mexican borders. Jonah yelled, "WE CAN'T FAIL NOW!" and then looked calm as he looked up at Titan.

"There has to be something we can do?" Titan asked.

"Yes, we have many cards to play yet." He turned and started calling out names. Soon people were following him into the transport room with the small shuttle pods, Titan noticed there were three pods waiting with doors open. "Two per pod. Let's hurry people," Jonah was saying as Ti entered the third one with his sister. It didn't take long before fifteen people were all standing within a dark room. Jonah was busy typing passwords in an ancient looking computer console. He pressed the last button and started speaking as lights filled the room.

"This was the place where most of the first prototypes have been produced. We are basically making super weapons down here... I remember my grandfather was so proud with the first design."

Most didn't realize what they were looking at because it was so weird and they were standing on a large metallic dish. Titan's eyes got big and he said,

"This can't be the weapon."

Jonah answered, "No, it was one of many which eventually grew into bigger, and better. We probably would have cracked the earth's crust with the bomb we used, if it was down here with no shielding. Anyways, moving on, this thing is an ion generator in its core which we will use to start these badass mamas." They had been following him across the room and entered to another more modern computer panel- the room that had denied them. In a couple seconds things started to move. Walls gave way, which exposed fat looking rockets. "Give me a few seconds and these things will be weapons ready to go. The remote panel is through that door you can barely see. Everyone go in and take a seat. I will get these ready."

They entered a command room through a small hallway. Titan sat down next to Luna, he felt like it was déjà vu. Except the technology seemed older and his sister was actually next to him. They looked at each other with little expression. He was glad this one was remote controlled, and it felt like they were safe- deep in a bunker.

Their screens came on abruptly, and with so much chaos, they had no idea what was going on. A Commander was giving orders to troops who were shooting and wounded in a brutal firefight. Communications were overwhelming with cameras on the soldiers and surveillance cameras as witness to the fighting.

Mexican battleships destroyed the five globes and as if ghosts would rise, did not stop bombing the shells of the spheres. A small stealth satellite had been picking up feeds and coordinates in Alaska and around the world for the last ten minutes after the attack. They had been graphed onto a map with many different targets. The Alaskan laser was target number one on the list.

A screen came on with the face of Jonah. "Okay, are we all ready? These things are very powerful, so have fun. This could go on for awhile."

They felt a rumbling movement, and Titan had a cockpit view of his craft. They heard and felt a thump and an explosion as the craft's rocket fuel burst into ignition. The rocket leap out of the ground and careened through the city and into the sky. The rocket exploded but only to reveal a sleek little ball which looked like a large oval, black with a deathly eye. It was small but still twenty feet in diameter and was winding itself up. The targeting took a lock on the laser in Alaska hundreds of miles away, and Titan read green... he pulled the trigger. It took a second before the craft shot its cannon. Not that it mattered, the targeting systems of Terra were astronomical calculations, calculating speed, curves and distance long before they happened.

This weapon was a rail gun which shot a small, inch-sized BB at a very tremendous speed. It was followed by a trigger which detonated the volatile pellet at its designated target. It was thought that many of these atomic pellets shot at a rapid fire machine-gun pace, accompanied by many cannons firing at once would cause enough stress on one side of the meteor. This would cause implosions and movement but probably not enough for such a nasty object in the sky. On the other hand Terra had created one of the world's deadliest projectile weapons in an almost perfect form.

The bullet was detonated milliseconds before it hit its target. The small BB was a heavy radioactive metal and atomically unstable. The speed and force of the growing bubble was a shockwave to the area, crushing trees to splinters and throwing the tower to the ground like a small flame to a wind. If matter felt the immediate shock, then it was no more, as things vaporized in its rapid explosion. The quick propulsion of the craft turned 180 degrees and locked onto the Mexican battleships. Titan saw the green light and the ships through the scope and quickly pressed the trigger. They sank as quickly as his view moved. By now his sister had gained control of her own weapon craft and had begun firing on American and Mexican weaponry.

Titan's view changed and he was looking into the stars and directly at a satellite with a big missile which had already launched. Its targeting showed green and he tried to adjust so that it would destroy both the missile and the satellite at the same time. It worked, and he moved onto the next, and the next, and the next and he wondered how dead the rest of the world really wanted them.

Luna took precaution and tried to destroy enemies which were surrounding New Terra. She was terribly dismayed at the destruction which was presented before her, and the willingness at which she participated. She could not believe her eyes from the massive devastation created by her weapon. Eventually she could not even look after pulling the trigger. Her hand started to move by itself while she could not believe that she had just become a murderer.

Titan on the other hand felt at home, he felt patriotic and noble and he knew he was helping the world. The small craft he maneuvered was technically perfect... he imagined. It seemed to glide so easily through the atmosphere, and it accomplished its task perfectly. It had its auto-targeting, but he still controlled it like it was natural. Time passed had seemed like half the day was gone and he realized his sister was not sitting next to him anymore. He had crossed the globe destroying anything that he and his targeting system thought could be a threat. This included hundreds of battleships, thousands of their aircraft, which launched with their sights set on New Terra. He figured he must have downed at least a thousand military satellites, scores of military trucks, tanks, and missile silos which were littered around the world. And he felt good about his patriotism, and the ability to destroy the world's archaic weaponry. He felt good, except that Luna was not next to him. His systems showed half empty energy storage tanks, and he had an empty tired feeling. So, he piloted his craft back to Terra from the Pacific, and docked in the bay which he was sitting next to. It took a moment to update, transfer his position and he walked into the bay which now seemed as busy as Terra's hangars. He felt at home, and he felt like he helped New Terra survive. Then he was witness at how terribly incredible and deadly his craft looked and he felt even better.

A cold snap came to his face, and he realized he was being stupid and maniacal. He should know better. Even this weaponry was old and weak compared with those massive globes. He was an ant compared to interstellar giants. He could be destroyed just like a bug on his foot. And where was his sister?

He found Jonah and told him that he needed to find his sister. Jonah seemed happy and tired and he said he would find someone to take over for awhile. He had seen Luna crossing the floor to the apartment entrance. Titan followed this trail and stopped only to look for Luna in Alisha's apartment. It was strange walking through the winding streets. Everybody felt patriotic, and although it was dark, there were many people who looked as if they had things to do.

He got a call, a small vibration on his wrac …"Hello?" Ti asked expecting to hear Luna's voice.

"YOU KILLED HER, DIDN'T YOU?"

Titan was shocked, and he realized it was his friend Marcus. "What do you mean I killed her?"

"SHE was trying to get asylum in Terra and something destroyed her transport. WHO was it, Titan?"

"I don't know, but you know that I don't want to hurt anybody."

"Maybe it was your sister, eh? Maybe you should fucking DIE." and Marcus hung-up.

Titan tried to call his sister and no one answered. He found a car drove like a lunatic and ran the rest of the way home. He yelled for her before he crossed the threshold into their home and immediately went to work.
"Clearance override eleven, I need surveillance. Find Alluana, and find Marcus ."

Location of Alluana is ... wrist access computer is at home. He looked over and sure enough her wrac was on its charging dock, and he heard a muffled cry. He went to the closed bathroom door and knocked. The shower was turned on, and Luna sounded like she was sobbing.

"Luna?"

"Go away…I don't want to see ANYONE."

Titan thought for a second and replied, "Okay."

He went back to the wall screen which showed a patrol car speeding through one of the easternmost gates, the cockpit camera showing an ashen, frowning Marcus. Dirt beaded in sweat down his face and along his suit, or it could be blood?

"Communication feed to that car please." *'Done'*

"Marcus, where are you going?"

"Fuck you, Titan. I'm the one on the front line taking bullets. Then when a transport which I helped set up tries to enter… then… then it fucking explodes."

"I was taking out satellites and battleships. I wasn't even looking at Terra."

"Right, I hear your clearance 12 now. How the fuck did that happen? Why don't you figure everything out?"

"I can stop your car, Marcus, please."

Laughter… "You were never as smart or as talented, asshole."

Marcus went busy pressing buttons, and Titan's security feed died. A surveillance camera switched on and Titan watched the small car fade into the desert. 'You will need every ounce of that talent to survive out there,' Titan thought.

He brought up command feeds and weapon craft statistics. By now 20 or more of those craft were patrolling the skies at any given moment. He brought up the back-log of recorded events. It didn't take long and he matched the weapon crafts patrol pattern with surveillance around Terra. Found the crafts ID and pilot ID and brought up the info. He was taken aback as jaw falling. His hands twitched, pulsing as he rechecked the data. 'How could this happen?' "Why would targeting allow this shot?" Titan muttered.

It was Alluana. She must have mistakenly pulled the trigger... targeting was not updated. 'Damn,' he thought, he should have known she wasn't the one for the dirty things of war. 'How did targeting, how did he allow this mistake?' Titan was confused and dumbfounded. He had to look at the screen name once more, and once more after that. He was rubbing his face when he felt the presence of someone behind him.

Turning he found his sister staring at the same screen.

She was naked... wet, though swollen eyes

streaming shining crystalline rivers.

Her hands were shaking nails reached her head…
sinking into her temples… blood boiling to the
surface.

Her eyes rolled back into her head as her hands
raked downward.
It took all of Titans strength to stop his sister from
harming herself.

WW-?

The war around New Terra ended quickly. They
destroyed most of the world's armament within a
few weeks. The United States was in shock for they
were never accustomed to such a heavy military loss.
Mexico had been even more at a loss as Terra had
declared the land through the Baja into the Pacific as
new ground. New Terra also set up a new border
another ten miles out into the desert, and twenty
miles around the reservoirs. These reservoirs were
modern marvels and Terra was still the only place in
the world to take salt water and make it fresh for a
hundred percent of its population. Many places
around the world still had little to no running fresh
water. This had just gotten worse as many people
just gave up their jobs and quit. Or at least gave up
their jobs for 'vacations' or whatever the human
mind could dream of. Many went into binges of

alcohol and lawlessness as the depressed, corrupt governments wasted away. Time moved slowly as this change was not welcome and many lost all faith in human existence.

This was in just half a year, and within another three months, the two asteroids of destruction would try to annihilate each other. Many thought it would do nothing. Some imagined it would deflect the giant enough to save earth. And some imagined it would create an explosion which would destroy the heavens anyways. In any case, New Terra had been one of only a few countries to ally and prosper. Many countries declared war, or civil war was created within. Most of these were being run or funded by illegal activity. Police forces were becoming weak and even the most diehard officers were getting tired from constant work, realizing new systems were taking effect. Many police had to sleep at work and many only answered the worst of calls. With no great military force to help 'govern' and weakening police forces, the structures of law were crumbling. The weakening UN regulations began to fall just as easily as the U.S. Constitution. A depressed, bankrupt world had been transformed into lawlessness.

Everybody had felt this pressure and much of this pressure had been put onto Terra's citizens. Luna had been one who felt the bitter depression for the first time in her life. She tore herself down because she was lazy when she shouldn't have been. Her targeting had passed over the wrong place at the wrong time. The fact that friendly fire was an unfortunate side-effect of war was one thing which she could not handle. She had seen what had happened, and had even focused on Marcus, and the silent screams for his loved one had broken her heart. She had truly fallen into a dark place which she could not escape. She had hated herself and she wanted to take her own life. Titan was constantly worried and had not even seen her for weeks as she was being counseled by psychologists in the very best medical offices. Titan did stay busy at his work, monitoring New Terra, and any threat was now taken very seriously. He nearly became a commander whom they called a clearance 13. He did not know if he wanted to be a commander. He was already more distant from his sister than he had ever been. He had rarely seen her outside of the apartment. A few times at the office for dinner as he stayed busy and he slept mostly at the base.

He did not like how things had changed.

And time moves forward, faster for some reason.

Explosive immortality

The glassy meteor was not very big. Its weight however made it look enormous. The distortion of light around its body screamed power. And its speed was something which humans could barely comprehend. It was as if it was being pulled from some unknown source. But it did feel its antithesis before they would collide. It felt the forward projection waves before the heavy anti-material was sucked into its unforgiving belly. This material was the largest quantity of antimatter which had ever been produced synthetically. The meteor was actually more closely related to antimatter than any other definable matter. These two beasts would still create a colossal explosion? The light in front of the bomb was being deflected by the meteor. Deflection created an intense corona in the sky. The corona looked like a small sun which lasted half a day. Many watching thought that it had already failed when it faded. The implosion of light took many by surprise. An event of this magnitude had never been recorded. Fortunately the Sun's heliosphere within the solar system helped deflect most of the intense light from the cataclysmic explosion. Auroras lasted for half a day and most astronomers were dumbfounded as they could not locate the remnants of the meteor. The last thing which it should have done was completely disappear.

Chapter Two

Boundaries

Marcus did not stop driving until his cars battery died. He sat silent for awhile but his rage did not subside. He felt like an outcast from in his home country. New Terra demanded hard work from its citizens and he always felt like a slave for the promotion which had never come. He felt like he worked harder than anybody, and even had many brilliant ideas which were always taken with no reward. At one point he decided to reward himself with alcoholic binging at the Acropolis night clubs. This only worsened a depression which he never showed. But a day had finally come with a new reward. He met his American love with whom a new light and meaning for life had blossomed before his eyes. They were like magnets for one another- true soul mates. He started to dream of a family with his love, and had forgotten about the meaningless promotions. She was even accepted into citizenship and with haste because he pushed for it. Now Marcus would only feel his heart beat when he imagined her smile, along with her gorgeous eyes that would radiate all the love in the world. He imagined those beautiful glowing eyes and he burst into sorrow. Tears streaming onto his jumpsuit and his fists clash with the interior of his car. He could not even hear his own screaming for his delirium destroyed his own senses. Eventually he passed into a silent void which seemed to be his only refuge from himself.

The window smashes inward and Marcus wakes with many tearing hands pulling and binding him into restraints. Marcus was thrown to the ground and he felt many feet and fists landing into his flesh. The pain was nearly non existent which was exactly how he felt. He wanted to die and he started screaming.

"KILL ME! ... KILL ME! ... KILL MEE! ... KIIIIILL MEEEEE!"

The pounding stopped and he slowly wormed his way to his knees.

Marcus spit blood, something white hit the ground. "What the FUCK are you waiting for you FUCKING LOSERS... KILL ME!"

A red woman approached, and he could barely make out what she said but he thought it sounded like "maybe". The woman crouched and grabbed his already swollen face. She looked into Marcus's eyes while she produced a syringe which found his neck and the darkness for him was most welcoming.

Marcus woke and blinking swollen eyes he realized he was not blind but it was dark out. He was freezing for the first time in his life but it felt nice against his naked, swollen body. A deep moan sprang from his lips as he shifted a stiff and painful body. Marcus realized his eyes were nearly swollen shut and he tried to open them but found that the dim light was worse than the dark. He did not care about his situation and he wondered and hoped that he would soon see his only love. A whimpering sound soon turned into screams and fits of rage. This brought chills down the spines of hardened men who were on guard in this underground lair. It made them wonder if he was an actual Terranian citizen or just a criminal outcast like most. Within a few minutes Marcus passed back into his dark peaceful void.

Awakening from the cool void he took his anger in stride. He did not open his eyes because he wanted action- and destruction. He figured that it must have been weeks and that maybe he was in a self induced coma. His brilliant mind had gotten the better of himself and he knew that he must keep his cool if he was going to survive. Coherent contemplation started invading an active mind and was soon thinking of a world with no technology- how great would that have been. How Adam and Eve must have lived like gods among creatures. No pressure other than to eat and live in a beautiful flourishing world. Creatures would coincide in a string of lives all helpful to one another. The death in this world should only bring more life in the proper ecosystem he thought. How greedy human nature had brought nothing but death and destruction, and how technology was just a tool for this. His resolve was becoming clear- to destroy New Terra and any other form of organized technology. Only then would he see his true foundations and only then would he resolve this nothing of a world. His fists clench and he felt his restraints tighten. He tried to feel his numb body without moving and surprisingly it felt better. His assumed that it was medic's office with an artificially induced coma and maybe he got used to the drug- he was dizzy and almost spinning. Either way he knew that he should keep cool until somebody cleaned him up. He cracked his eye lids enough to let the light in. When his eyes had focused and were nearly dry he moved them back and forth. It was definitely a small room and most likely a medic's office. He saw the bag of saline which was nearly empty and knew that a camera was now probably monitoring him. He shut his eyes and again felt his restraints. The tightest one he chose and began slowly and methodically twisting his wrist. It took what seemed like hours before his tender skin gave way and produced the wetness of blood. During this time he figured that it must be night or close to morning and that his surveillance was probably lax or nonexistent. He felt for restraint around his legs and determined that there was none. He would be easier to clean that way and it would also help his escape. The slow twist of his wrist stopped when he felt like the blood would be adequately visible. Marcus relieved his bodily fluids and prepared himself for his actions. His mind played out his every move, and every twitch of his muscle. He did this for hours while he placed himself in a meditative state.

The thump of sound woke Marcus and he placed himself on red alert. This is what he had been waiting for and he was not going to mess this up. The unlocking of the door is what had tuned him in, and he heard two sets of feet walk through. It did not sway his resolve at all.

A woman's voice broke the silence. "Let's do this quick, I don't care about this asshole and if he really is Terranian I don't even know why her majesty wants him alive."

A man's voice replied, "He is Terranian because of that suit, but he stinks like any pig. Where's the air freshener?"

"Please, it's in one of those cabinets above the sink... Aw his wrist is bleeding I told you not to tighten these too tight."

"Good, fuck this guy, hopefully it's infected."

"Well, let's see, might have to cut it off. We got time for that, right? And could you get the damn freshener?"

"Yeah babe, we got time for that."

The man started to rummage through the cabinets while the woman smacked Marcus in the face a couple of times. His limp body showed no sign of awareness and she turned her focus to the bloody restraint. While the man sprayed air freshener the woman unlatched the tight belt. A loud Velcro sound ripped the air and a gale force motion blurred across the woman's lazy eyes. Marcus was so fast she did not even have time to gasp for air before an iron grip landed on her throat. This motion landed like an uppercut beneath her jaw and nearly lifted her off her feet while being twisted onto Marcus's lap.

"I will rip her FUCKING throat out if you try anything. Don't even move towards me... And you... untie me or you die."

The man was putting the air freshener away when this had happened and he turned slowly while Marcus spoke with a look of dreadful intent. The man could see a sinewy arm which released enough for the woman to take a breath. Marcus looked at the man with cold eyes while the woman fumbled with the other restraint. Color returned to the man's face and Marcus knew that he was thinking of something.

"Now, take out the needle, and if either of you even breathe, I will break her neck."

She finished her job and Marcus moved with a weightless grace that would make a ballerina smile. He went after the man and crossed the room with the woman still in his grasp.

Marcus took his time when he realized that no surveillance was in the room. 'An even playing field,' he thought, 'how nice.' He did not kill them but took the man's clothes and had broken the man's nose in his frenzied may lay. Now the man lay quiet on the bed which was still warm from Marcus. He sat the woman next to the entrance and gave both a massive dose of the drug which kept Marcus unconscious. He knew he must move before somebody expected these people for whatever they might have scheduled. He found himself in a long hallway and he figured he must be in some kind of medical facility. He heard voices and moved in the opposite direction. Hallways emptying into a weaving exit towards a flight of stairs and a quick descent into a parking garage. It was mostly empty except for a few cars and a dozen or so motorcycles. Marcus picked the one which resembled more of a dirt-bike. Tearing into the wiring all he had to do was get ignition to ground for a few seconds. Cursing under hot breath as a man entered from a different staircase. This man was oblivious as Marcus crouched behind the bike. He turned on his own in a sputter of freedom, and looked at the ceiling of the garage as if he was glad to get away for the day. Marcus crept like a stalking panther and drove a needle into the man's neck which made him slither to the ground as a limp noodle. He took the helmet that the man was holding and peeled the bike out of the exit ramps. A city which had turned into what looked like a military base passed before Marcus's eyes. The streets were deserted except for the occasional motorcycle or roadblock. This he found strange as if it was funneling him towards his only exit. A siren sounded as he found relief with what looked like his last escape route. This road he realized was going north into what looked like a small canyon. It's a given path and his only choice so he gunned his bike at full throttle.

Twenty minutes later the small canyon disappeared and Marcus finally relaxed because nobody was following him. A strange feeling crept over him like he was being watched but after slowing down, peering into the desolate surroundings this feeling vanished. Hunger shot through an empty stomach and Marcus wondered how long it had been since he last ate a solid meal. This he pushed aside as he focused on the road and its endless pavement. He hadn't even bothered to get his bearings but assumed that he was going north, or maybe it was the gradual ascent of the land. In any case the winding road soon crept over a hill which presented a small town through the haze of the heat. Marcus was pleased enough to talk to himself, "Finally, hopefully food and a sane bed." His hopes were withdrawn as a strange… seemingly deserted town crept into focus around the bend- closer and the road went through this. Windows were boarded and gates had chains with big pad locks.

A restaurant appeared to be open and his stomach grumbled in angst. It seemed deserted enough and he had a few bucks from the sleepers. He tried to compose himself and strode in as if nothing was wrong. People were despondent and he sat at the bar with the rest. The bartender appeared happy and he ordered a drink and snacks for food.

He ate fast without touching his beer, and when heads turned to gaze he drank.

The air was heavy... a ceiling fan buffeted and shook. Stale, dry, quiet... Marcus realized he was already finished and he downed his beer only to cough and choke.

"Hey your momma teach you to drink?" the closest patron asked with a smile.

He had just been put through hell and one person was not going to faze him at all. He asked for his bill and paid to get moving- he did not even know where.

Outside he nearly ran to his bike for a creepy feeling took hold. The bike churned igniting its purr and the glare of a snipers scope was the last thing to catch his eye as a thump hit him in the chest.

***

Marcus felt a very different waking this time. He woke to soft music on a soft, fuzzy bed. He rose slowly rubbing his aching head and at the same time realizing that he was naked. Looking around the room Marcus found and put on clothes. The room was expansive and his nose realized the massive table of food before his eyes did, one of the freshest most bountiful amounts he had ever seen without many people surrounding the table. He looked around once more before diving into the large pieces of meat and fresh fruit and vegetables. After a decent meal or two he noticed the bottles of wine and figured why not, something to take the pain away. A large glass was poured and he sunk into a plush couch. As he scanned the room he still could not make out the outline of a door. It did not disturb him at all and he wondered if he was finally dead. He drank heavily and felt tired until he heard a moan. It was a waking moan followed by a girlish giggle. He stood as a curtain was thrown aside to reveal a gorgeous full-figured woman. She was very dark in hair and eyes but his eyes were directed at her incredibly showing lingerie. She smiled and started walking towards him and stopped as her breasts pushed against his chest. She broke the silence as he was still awestruck.

"I have always wanted to meet a Terranian native, and you have been quite the animal. Most of my expectations have been met, but I am the Queen of Texas, and I need a King. Will you be mine Marcus?"

"I... I am no longer Terranian, my Queen."

A large smile crossed her face and she viciously pulled Marcus to her lips. She tore clothes and scratched his chest until he bled while she pushed him onto the bed. She was an opposite woman from his American love; she was very rough. When he closed his eyes he imagined the past- beautiful, kind, lovely eyes, which made him very angry. His anger was accepted with grateful screams which lasted for hours.

***

Marcus was amazed at how powerful this woman warlord was. Her name, Angelina, was quite acceptable, but it was her nickname that gave dire respect- Angel. The Angel was an international criminal and Angelina was in control of Mexico. Her father and mother used control through brutal torture, a nature which transcended easily into their daughter. They would not idiotically kill like the mafias in the past, but their torture game was not even talked about by their unfortunate victims. Many of these victims would only say that they would rather die, and most were amassed with mental scars worse than any cut or burn. Politically they would control the Mexican 'Democracy' with a shadow government. This soon pushed into Texas when Angelina had taken control. She had no problem gaining ground into an entirely different country especially after Terra's worldwide weapon extermination. Her empire was drawn from the northern most parts of Texas and down all the way into Southern America. She decided that to try and take all of the Americas at once would only be foolish at this point in time, but it was on her mind.

Marcus fell into her ways easily, a little too easy even in his mind, but their goal was the same. They wanted to take the world and mold it into whatever they would like. He, of course, had to take some time to dominate her generals and cohorts, but when they realized he was maniacal just like their Queen, they obeyed. She lost most of her army in the assault against New Terra, all of her Navy and Air Force, but still maintained hundreds of thousands of soldiers who would die for her at a moment's notice. She did not even care about her losses as it just made her more amazed at this country which she could never visit. Now this Queen had her prize, a prize which she loved and who would help her fulfill dark dreams.

Free captivity

Alluna had been cleared by her psychologists as mentally fit. Her work, however, was still in question as Security could not allow clearances and her ambassadorial duties were now expired. A few European countries had stayed in contact with Terra, but the only one which had absolved the war was Brazil. The other countries had become stubborn and did not like to lose at anything. Thousands of stockpiles had been destroyed, and many countries realized that they should not have tried to help the U.S. which put their militaries in the scope of New Terra. They came to the aid of America, and a few weeks later Terra had come for them, and pin point precisions massacred the slow archaic weaponry of the world.

Titan finished his night-time patrol with His weapon craft satellite, which had now become routine. Things were looking up for the first time in some time, in Titans mind anyways. Luna had become playful and was laughing like days of old. Titan could not wait to get home, but had stopped to get some breakfast. His feeling was strange and as he walked through the apartment door, he knew something was wrong. He ran through the apartment only to find her wrac on its charger with a message. After reading the message, he threw it against the wall and headed back to base.

-To my dearest brother,
  I love you more than the world, but I must leave my sorrow and my beautiful country in order to find something more. Please let me do this Titan, I love you. -Alluana

***

Alluana had written the message and in a way knew

that it was not necessary but wanted to let Titan know that she would be free. After placing it on the charger, she glanced around the apartment and did not want to leave her home, but also knew that something had to be done outside of its confines. She took her small bag of things and headed to the Acropolis. The last few days had been fun for her as she honed her many skills within its playgrounds. She was an excellent teacher and had instructed many children in the arts of certain sports, archery, climbing, word usage, grammar. The one thing which she had never gotten used to was scuba diving. While she loved to swim the thought of great confining depths had always scared her- not that she was claustrophobic she just always wanted to breathe... until her fight with depression. Now it was the children's turn to teach her how to become drown-proof, while swimming like a fish with nothing but a snorkel and flippers. She mastered it almost naturally- after a few mistakes, and was soon swimming the large tunnel-connected pools with human fish at her side. Her alibi was never seen by anybody, even those with prying eyes. She only hoped that the ocean would hide her body heat enough to disguise herself from the watchful eyes of her brother and the other weapon craft which patrolled the skies. She thought to herself how lucky she was to have her brother on grave shift for this, and how things seemed to be falling in place. She rented the mask, snorkel, and flippers to aid her escape and slipped out of the Acropolis and back to her commuter car. The drive to the Pacific coast was nice, and she found the party by the ocean which was also part of her guise. It was a hypocritical party for the new boundaries of the free world which was New Terra. She knew some people but mostly kept to herself while she ate and drank non-alcoholic drinks. When the party hit its peak, and while people had light heads from many cocktails, Luna took action. The timing was perfect as the sun had dropped and a fuzzy dusk illuminated the land. She slipped back to her car and made like she was driving back home, but stopped outside of the party after parking the car off the road some distance. She knew her brother would probably find her motives before any commuter-car locator would. The party was a perfect excuse for this reason, and it made her think of Marcus. He must know the very intricacies of the car's technologies if he was to get away so easily. Not that they didn't let him get away, but Terranians are not blatant killers. Luna made a wide berth back to the ocean, far enough to go around any security but not so far to be picked up by the roaming satellites in the sky. One Security officer nearly saw her with his roaming lights but was remiss to the sounds of the party. She exchanged her clothes for those in the bag which was her Security jumpsuit. These suits were all-around the most durable in hot or cool and harsh weather- tight enough that it almost looked like a wetsuit when wet, but still resistive and not nearly as slippery. She packed her things and for some reason took the time to bury them in the sand which she found slightly humorous, for now she was like a defected spy. She was a captive, emotionally and physically contained, and she had a deep feeling that something was waiting for her outside of her home. Luna took one last look at the glow of her country in the distance, and slipped into the dark, cold water. She swam beneath the water past the breakers and made her way north for the rest of the night. It was a hard, long swim which made her worry for the first time in what seemed like an age. She wondered if some hungry, nearly extinct shark would find her as a meal, or if a Terranian speedboat would come cruising in out of nowhere. Her trip was most uneventful and it made her feel lonely and small. She swam while her legs ached and burned, then she swam harder and faster. She swam past lights on the land past a depressing, repressed Tijuana, which had been cut off from the world. She swam until the breaking of dawn and decided to swim some more.

Titan was on the horn before he made it to his car. When they called back and said her location was at home he was furious, and when he started yelling, their job had gotten serious. She must have taken an id chip out of her wrac and used it to commute, so he told them to look for her last destination. When they informed him of the location about a half mile from the Pacific coast he realized how easily she must have escaped, and hoped she wouldn't have gotten far.

"I want a search team on the water, and I want a

rescue team on land, NOW!"

Titan made a turn at the next exit and started his way
north towards America. The next person to contact
him was the Flight Admiral of Security, Scott
Garner.

"Titan, we are doing all we can. We have boats on
the water and craft in the sky, but your sister made
an excellent escape. She joined an Ocean Pacific
Ceremony last night to celebrate our land
acquisitions, and she must have used the ocean to
make her escape. Most likely she headed north, but
Titan, I warn you, there are roadblocks and militias
on almost every border. We cannot afford a total
ground-assault into unknown territory for one
person."

"Damn it Scott, I cannot leave my sister out there
while the world eats her up. I have to try and help
her," Titan said this with clenched, quaking anger.

"I know… that's why you're free to leave if you like.
Stop by the northern Security border and pick up the
car that we will pack for you. And Titan… don't get
yourself killed." Garner hung up before Titan could
say anything else. He wasn't the only one who was
furious about this situation and he tried to drive
faster.

Titan went to the Security border and realized that it
used to be called the Security checkpoint. He

dismissed this and got inside his modified car which seemed gutted but armored among the modifications. He would have enjoyed it on a different occasion as it seemed like one of the James Bond cars from one of the films which he liked. After a quick debriefing about this car he sped out of Terra and towards his sister. Or at least what he hoped would be towards his sister. He could not believe that she would do this to him after she seemed like she was getting better. A holo-screen popped up which showed a topographic map of his surrounding area. Within just a few minutes he was to encounter a road block which was easily passed by going off-road. This done, however, someone was in the sentry post who most likely had contacted somebody, which meant this journey was going to get interesting very soon.

A new voice clicked on while driving through a town with many gawking people.

"Hello Titan, I am your eyes in the sky, and I can destroy any road blocks as long as there are no people around."

"Thanks, I will need you, what's your name?" Titan asked sternly.

"My name is William or Will, at your command."

"I won't be very pleasant right now, Will, but I am grateful. Have you found any location of my sister?"

"Not yet, we do know she went north as a small heat signature would indicate from a satellite last night, and she passed Tijuana, so she probably went all the way to San Diego."

"Then that's where I'm going. Keep me informed at all times, please."

"Can do... we have another roadblock ahead about five thousand meters, but they also have people in the sentry building. Want me to shock and awe?"

"Yes, I don't want any casualties... unless they mean harm," Titan said this with open eyes as he never had seen a rail burst up close.

"Ok hold onto your pants."

Titan saw the small building in the distance and a few seconds elapsed before a ground shaking explosion pummeled the ground. The ground quaked, closer to Titan than he expected as they did not want casualties. A shockwave blasted by and threw his car a couple of feet squealing off track, followed by a ballooning bubble of dirt and stone. Titan knew these weapons were incredible, but up close feeling the shock reminded him of just how easily things can disappear. He had no qualms while passing the now dusty outpost and he didn't see anybody and knew they were hiding, like their own disappearing act.

"That was good, but if we don't want to hurt anybody then we should probably go with a little more distance." Will said, but Titan knew distance meant towards himself, and too much closer and there would be no benefit.

Titan said nothing but thought that maybe he should have while his road was long and lonely, his head on a swivel through small communities. Hidden fear came on while on this silent desolate road and he started to drive faster.

"Bad news, Titan… It's still Will, but I am getting orders that I cannot harm anybody, no civilian casualties- no rail burst. We have what looks like the image of a swimmer coming out of the ocean just south of San Diego. She looks to have been captured, and after that, we cannot track her very well. She disappeared… into a subway."

Titan's heart dropped and it took a minute before he could place words into his own mouth. He could almost see the beach and he knew that he was close. The twelve or so hours of swimming would equal an hour or two of high speed driving. He tried to speak clearly.

"I WANT THERMALS, AND AUTO TERRAIN, I NEED TO KNOW HOTSPOTS AND HER MOST IMMEDIATE LOCATION … NOW!"

The sprawling city seemed to never end while awestruck people had been warned about an incoming threat. This did not stop people from exiting their homes to watch the small technologically advanced super-car rocketing through their neighborhoods. A small line of police cars followed which did not bode well for Titan. He knew that this could be martyrdom but his sister was more important to him. Soon he was close to the beacon which was the last spot that could have been his sister. He entered a subway and drove down the rails into the most likely direction of his beloved kin.

***

Alluana woke to the sound of an engine. She was moving on smooth road that she realized was some kind of transport. She was bound and while she explored her surroundings with her eyes she tried to remember what happened. People had greeted and quickly abducted her. She fought hard but they were well equipped as if they had practice. She realized maybe a day passed, and noticed that she was in the middle of a semi-trailers transport which was carrying cattle and chickens. This was the perfect transport for humans while her country's satellites looked for heat signatures which would be her trail. Time passed and the shipping of goods was her perfect transport. A second stop, a third, and she is on her way towards her final destination. Her memory kept giving way to brief events which ended in capture and on this road. 'Stupid,' she confessed to herself, as time elapsed. Five or six hours had gone by and the stop finally came. She imagined terrible things on her journey and only assumed that it would be a true payment for her sins. The light was bright and she was correct in her assumption that it was still only midday. She focused on her past hysteria and her pain which she had felt through the long swim and a sense of accomplishment was twisted in her mind. Luna didn't mind being pulled out of the transport and was pushed softly into a very large building. She felt absolved and realized how many people were standing around that were holding on so gently. It came as a shock that they actually seemed interested and amazed at her 'foreign nature' while they descended a broad stairwell. This was not what she had schooled herself to expect but it was welcoming and she started to cry. The people gazed at her in an even more astonished tone with awestruck eyes. They did not push her anymore but instead just pointed down the now apparent hall which had been presented before the large throng of faces. She held back tears before her descent into whatever may wait for her at the end of her journey. She still did not expect anything good, but after walking for sometime, people had diminished in their downward tier of hierarchy. Soon she was walking down empty halls which lead toward what she left to find, crossing into a warm room with a burning fire which seemed to light a picture from an age long ago. Pillows and furniture adorned a seemingly enormous, unending room and for a minute Luna had thought that nobody else was present. A long, broad, inviting table held the many foods. She began to eat and realized just how tired and achy her body was from her exile. Eventually she found the end of the table which was full of different alcoholic beverages. She regretfully declined what seemed like an open invitation and soon after heard laughter. It boomed in deep strength. It was coming from the fireplaces corner. A large creature in a black shadow moved and stood. He turned and presented himself as Luna was awestruck. This man was incredibly tall. She tried to compare him to her brother and found that to be impossible as her brother would have to look up to this man. He spoke with a melodious low tone.

"Many people are quick to drink such fine liquors. I must congratulate you for that, and I must ask why you cry. Do you have great knowledge?"

"I … don't know, they say I can't dwell on the past, that I can't change anything." Luna felt very calm in what felt like a shadow of his towering stature and she realized why the room was so big.

"They call me Jesus, but my birth name was John. Will a foreign woman with much on her mind talk to me about her past?"

"Yes, do you have some water?" Luna asked while she tried to look into the big glowing eyes.

It was this giant man's turn to be surprised and his

large smile opened into a rumbling laughter. While he was still smiling he moved to a large container which looked like it also watered all of the many plants through tubes along the walls. He poured two large containers and motioned Luna to walk with him to the other side of the room. Luna now noticed the den which stood out as a large corner and realized the room might be in the shape of a colossal cross.

"Nobody asks me for anything, they always give." His voice took an eloquent roll. And he noticed her gazing at the intricate stained glass which glowed with sunlight.
"Yes, it's impressive, but I figured that things needed to be changed for the better. You are in a temple in Utah; my people were Mormons who now abide my teachings. They believe I created the satellites which destroyed the world's threats. They believe I also am the creator of this worlds blessing. I admit that I had to change things in our culture and many people are better for it... Will you tell me about your past and your country?"

Luna took a gulp from the large glass and started to explain her position in clear detail. She began by saying that her country was full of helpful willing people who work for one another. No great excess is taken because the country is quite small but made of vast towering cities. How until recently, nearly 95% of New Terra's ground had been terra-formed in some way. She exclaimed that she is very close to her brother until recently. She started to cry during this time while John sat still and patient. They were still close but she could not take warfare where her brother thought that it was patriotic, and justifiably would be self-preserving, and beneficial for mankind. She told of how she tried to find solace after accidentally killing a friend of hers, but it was the confinement which clung to her and made her want to escape. Quickly going on to say that much of the technologies which had saved mankind had been kept very secret, and they were basically massive weapons. Not knowing herself until things actually happened, but that she lived a better life knowing nothing of a possible doom.

After what seemed like a long thoughtful explanation she sat with a still, wet, puffy face, but she smiled and for some reason felt better. She sat back and relaxed and asked a rare question to John.

"You must be a good man, and I bet you like to be called John. Do you think that you are Jesus?"

John smiled his wide beam and spoke. *"You are right, I like my name John, and no I don't feel I am that great. I bleed like a man and have headaches which are from my size. And the people think that it's from my invisible crown of thorns. They will still get me aspirin, they get me anything."* John bowed his head and explained his position a little more clearly to Luna. He told of how his orphanage ended strangely at the right time and place. How a great maniacal minister had died after his arrival. A giant baby who appeared out of nowhere had slain a great sinner. He explained growing up in the most sheltered and gracious lifestyle. He never knew why everybody thought of him as special until he recited a passage from one of the Bibles. He learned how to read through the Bible but this was a strange incident at Mass. Before he was ten he had been sitting in the front while the preacher had almost started his service. He had reached down and picked the Bible up, turned to the middle of the book and read one line which made everybody bow. Even the priest was bowing and he felt very strange so he kept reading while people sat and listened. Soon he was the preacher, and for awhile had enjoyed it until he realized how things where run. Many husbands would beat their wives and slavery was nearly legalized as contract labor held people in bondage and poverty. He explained how things happen with haste when he gets upset, and that it could be considered power, but not like the kind which is found in one of the smallest countries that is New Terra. He then asked a question that Luna did not expect.

"I would like to get away as well. Could I someday visit your country?" He said this with big, hopeful eyes, the likes of which made Luna's heart weak. She felt absolved of her sins and gave the big man a hug.

"I need to tell you something else…"

Pain

Titan drove for ten minutes in the subway tunnel and got annoyed. His systems were giving estimates, but communications had been severed, and he had no idea if he was even close. He turned around and started driving the other way. Ten minutes later he was becoming desperate and started to scream in annoyed anger. The only thing that stopped him was a dim light which seemed like the end of the tunnel. He slowed as the light got bright and he couldn't see what was blinding him. It looked like the subways train and Titan could not tell if anybody was inside. It seemed like a strange thing to have the lights on with nobody around. It would not be good if this train started. But maybe they wanted to tell him something. A minute later Titan decided to get out of the car even though it felt wrong. Maybe somebody lived down here, and maybe he could ask them where he was. He emerged from his car and walked to the train's cockpit. Nobody was inside but he heard movement from behind. He turned while being struck in the back of the head with a thump. Titan fell to his knees and the attacker landed on his back with legs wrapping around torso, and the cloth was already on his mouth as his breath was deep and his tired mind went blank.

Titan woke and decided to listen and stay quiet. He was being traded through some kind of underground black-market. People were arguing but it sounded like things were calming down, and Titan felt trapped and stupid. They said what sounded like "You get more than just money" and they made the deal. What an idiotic position he had put himself into. He knew that small factions had taken over but did not think that it would be so close to home. He had been loaded into some underground train. They started to move which was surprisingly fast and smooth, but in a cage like an animal and his hands were bound to the bars with plastic shipping ties. Almost like they wanted to give him a chance if he tried but he did not know how strong the cage was or if he could open it at all. They must be smart to be doing this under the nose of powerful satellites. They were at least 100 feet below ground, and on what seemed like a very well kept rail system.

The car slowed and made a sharp turn before another lengthy ride. During this time Titan realized that he was not the only cargo, and that people where talking close by. He was still tired from his night shift; the nap was welcoming and the sound droning with a consistent thump of the rail car that would help him sleep.

He woke when the car stopped and the tunnel gave way to a large shipping room. Small forklifts and trucks were waiting to be packed and driven. The people got off the train and treated him like cargo. He realized they did not care if he was awake or not. They were professionals and would probably just shoot him with tranquilizer if he made any commotion or noise. He'd been placed at the end of the shipping train, in a large steel box, between two heavy plates sandwiching many thick bars. Eventually after a few armed people packed the front of the transport around Titan's cage a most unwelcoming container was placed next to him. A steel cage similar to his own held a giant panther pacing in some trancelike fury. The forklift driver made it slam next to Titans cage and the scream that bellowed would freeze the strongest of men. A paw entered Titans cage by a few feet and deadly claws were produced and retracted on screeching steel. A man spoke to Titan, before kicking the opposite side of the panther's bars.

"She likes exotic animals."

The antagonist made the giant roar with a deafening note. After what seemed an eternity of spent air it started its pace again and the truck moved. He was a living commodity for someone and was packed in a large armored train- with a beastly feline companion. They began their journey, the plastic wrist ties were easy enough to break, though strong enough to make the wrists bleed. He felt his pockets which had been stripped, unzipped his suit, and felt for his own tiny pocket with his father's small pocket knife. It was found and it made him feel slightly better about his position, but not by much. He realized the boxes were gated at one end and probably had a padlock on the top of the box. He was in the box of an animal being stared down by a hungry beast. Its gaze was dark and full of blood thirsty hatred. Even if he was fast with his knife he probably would not win a fight and hopefully, whoever she was, hopefully, she was having a good day. He pulled on the gate and tried to open it but only ended up angering both himself and his trapped beastly passenger. He sat with crossed legs and tried to remember his meditation classes. Placing himself in calm realms in mind and Titan decided to try and hum a soothing sound for the beast. At first it did not seem like it enjoyed this noise, so he tried a different more baritone melody and found a quiet stillness. The panther sat and stared at him with dark questioning eyes. Its gaze seemed intelligent like it knew that they were both in the same position. It smelled the air with its great nostrils and made a huffing, growling noise. Titan closed his eyes and focused on his deep hum.

Eventually this path came to a stop but only to rest for the night. The drivers stopped at what must have been some kind of transport depot and motel. In the morning they were handed food and water. Titan was given a bag of nuts while the panther got a big juicy steak. They handed Titan two bottles of water and told him that he could water the cat. They laughed, and started the train on the way. Titan poked a hole in the top of one of the bottles with his pocket knife and moved the bottle slowly to squirt a misty stream into the cage of the cat. It growled with a vibrating pulse which brought chills to the back of Titan's neck. The beast smelled the liquid as Titan took a drink of his own water. He leaned his head back against the cage and asked his sister why she had done it, and he prayed, and he hoped that she was all-right. The giant beast sniffed the air next to Titan's cage, and Ti slowly raised the bottle and put a small puddle of water on the cage where he could. It smelled and licked at the puddle and its stream of water. It did this for a minute to get a taste of the refreshing water. It must not have had any for some time. This made the great beast pace, but after some time, they both fell asleep. A day must have passed and they made their destination as Titan and the great cat woke before their halting screeching train. The great cat began to growl, and Titan heard commotion and laughter. A woman gave orders, and workers fell quiet as a lift truck dropped its chain onto the top of the cage. The beast made a deafening roar. Its muscles rippled, and Titan was witness to its exquisite primal features. Large razor claws were feeling the ground waiting for the moment to pounce. The woman's eyes were big and she stared with the same delight when she looked at Titan. She spoke with a stern, demanding nature.

"Unload them! You all know where."

A servant asked. "Do you want him sedated?"

"Yes," she replied casually "Not now though, not if

he doesn't scream- until I make him." She looked at Titan with cold eyes which he found himself relating to the panther's.

His final stop was in a dark prison-like room in the middle of some kind of urban compound. This room had an operating table with a light that looked big like a doctors. The large, shiny tool-box gave Titan the chills.

***

Two people were standing at the door when he woke. He was lying on a table in the middle of the bright light, strapped to it with arms stretched. Wondering how and in what world could make this happen so easily- in what he now knew was no enlightened age. What did his sister know which he did not, and how would he get out of this? His arms were strapped firmly to the table and he had been stripped of his jumpsuit. Prisoner of War... and he thought it was fitting... how much pain has he caused for the survival of the species. He shook his head and told himself to not imagine death, just yet... no matter what the cost.

The two were not who he expected... a beautiful woman who looked like she could be a model... and Marcus. Titan could not find words and Marcus spoke quickly.

"Look at this beautiful position, ah Titan ... ha... never got that name. What your parents think you would do anything for this world? No, you were given the role which you played, and you played it well. Our country may have saved the world, and we are the fodder- the born killers."

"Marcus... things should have been different." Titan bellowed back.

"BUT IT'S NOT!" Markus yelled, and then spoke more calmly. "You're human Titan, nothing special; you have done your job well, but it's time for revolution. I have your car, wrac files only a matter of time and we will hack the satellites."

"Marcus you know they change codes constantly and my keys will not work with me gone."

"So, maybe... I'm lucky that the doctor was away while I fell into beautiful grace."
And Marcus kissed the woman and they exited the room.

Titan yelled after him, "Love should not be reserved for those living!"

The door closed and he wondered if they even heard him. He noticed the cameras and knew that he was being watched for this. He could not tell them any codes... expel them from mind.

The doctor entered.

"This will be the warm up stage, no questions."

The doctor was old and dark, even in his motions. He pulled a machine out of a closet with a boom arm and attached to that arm was a syringe. He moved the machine into position and swung it over and onto his hand while Titan realized each individual finger was strapped down.

"Don't worry, you won't get used to it, and screaming gets old. Just tell them what they want."

He didn't wait for a reply. The doctor took his time and put in earplugs, emotionlessly pressed a button. A needle was thrust down into Titan's index finger. He winced as a fiery pain shot through his finger and down his arm. It felt like something crushing his finger with liquid fire and Titan remembered his fight with the first man he killed. A second needle pierced Titan's middle finger. Eyes bulged and he remembered the dead spy's own bulging eyes as he had driven his knife up and into the man's heart. A third and forth needle brought spasms of pain. They must have been laced with some kind of nerve poison. His one hand throbbed and felt a deep, sensational pain which could only be subdued by deep breaths and meditation. The last needle felt like it pierced through the bone into the thumb, and Titan's eyes rolled into the back of his head.
He was soon wakened with a strong sulfur smell...

and a melting hand. The man was moving to the other side and Titan looked at his hand which did not look bad at all except for swelling around needle points and incredible pain. The machine was placed on his hand and tears sprang from Titan's eyes while his head rocked back and forth, waiting for mutilation.

***

John told Alluna of the southern devil, the name simply known as Angel, who would use torture to control, and underground highways to command. How his men had spotted Luna first but any daytime activity around Terra would be noticed by 'Angels' army. This made her wonder about her brother and she had only one question.

"Was anybody abducted?"

"Yes."

She knew who it must have been and also what John was doing before he said it. He was building an army to defeat this 'blasphemer', a religious army who would die for causes greater than themselves. John explained that they had committed strange murders recently, and had nailed men in trees. This created a stir and it was one that John exaggerated. Many generations of religious migrants had come out of hiding for their lord, mostly to resolve their own sins. He told them that this was the final battle and test which would determine salvation, and they must fight for their land in paradise. He had many factions all dedicated to him at a moment's notice. Luna told him how he was more powerful than any citizen in Terra could ever be, and he replied that that is not a good thing.

The Black Widow

Angelina could get virtually anything she wanted at a moments notice. This made her very jealous when she would want something which could not be hers. It was this driving force which made the world her own in her mind. Everything in the world should be hers, and things were falling into place. But she did hear what Titan said before they left. She did not mention it to Marcus who may or may not have heard but she was going to get answers from Titan anyways. His first taste in neuro-chemical pain had been given. A few hours had passed to give them time to think, and now it was Angelina's turn.

She entered the room and smiled. Titan looked at her with eyes full of hatred and could not find words to ask why or how. She started to ask questions.

"Do you know who I am? I am the Queen of this world and will soon have it all. And I have questions, Titan."

She started by giving him the same treatment. All of his fingers pierced by neuro-chemical toxins. He did not scream but instead moaned and wheezed. He did not care about telling her of Marcus's lost love with whom his own sister had killed. Titan even told of his life and the lifestyle of New Terra which Angelina did not seem to care about. After she was done with his hands, she moved the machine down to his foot. His hands and arms twanging in hot pain, and he could not imagine the pain about to pulse through his legs. She asked the question and pressed a button.

"What are your codes?"

It felt like breaking the heel of the foot as a needle was driven into the base. Titan tried to twist out of the restraint which barely moved. It would also hurt more as the searing poison inflamed tissue and nerves. He screamed in pain.

"THE CODES?" She demanded, "The ones which control Terranian weaponry- you know. TELL ME!!"

She hit another button and the top half of his foot felt as if it had been severed with burning fire. The fire moved up Titan's torso and he thrashed in pain, while he tried to hold his burning leg still. She screamed. "GIVE ME THE CODES!"

He remembered coding from a programming class long ago and he started to ramble numbers, another burning needle would insert its hot agony. He screamed a loud painful twist which he did not know he was capable of. She laughed, and asked her questions. While she asked he did not listen, and barely remembered the one blinding code which he could not make himself remember. Blinding pain followed, and the dark code tried to creep, its black ink onto the paper. More blinding pain... a beating pulse was felt and his heart felt like it nearly wanted to stop its thumping rhythm. He could hear it in his ears and he focused on the white brilliance that cast its warming gaze in his pain. He felt like his entire body was on fire and everything seemed drowned out from the real world. Crispy flesh was burning from the inside out and he felt all his organs in wavering agony. Melting, twisting- bones would twinge in the throbbing fire, as he heard ferocious screaming. He could not tell if it was his or hers and he decided to pass out.

Waking to find hands around him while they un-strapped and put him into the cage. His mind did not even want to remember all of the single questions which brought twanging, knotting muscles. He passed out in his cold hard confinement curled into an agonized ball.

The next few hours brought the same feeling. Not at first, they let him eat and clean up, before they sedated him and strapped him back down. He actually didn't feel bad until they brought the nasty machine out on its wheels. He could hear its squeaking… moving closer. Three people entered and Titan knew this would not be good. Marcus spoke first.

"We need those codes… just give them to me and we will not bring you anymore pain. We know that time is up. Your codes would probably shut down enough security for a small squad to get into Terra. Your time is about up, Titan."

Angelina spoke to Marcus and the doctor.

"Leave, Marcus. Let the Doctor do his work… Doctor get busy." Marcus left and she said one more thing to Titan before the doctor started.

"If you survive this then I might let you go."

Titan did not know if he could.

***

Luna made contacts with Terra as subtlety as she could. Through an underground system she notified them of 'Angels' growing army, and their security risks in Titan's position. New Terra was surrounded by a gangland, more dangerous than any perceptible government activity. She was terribly worried, but soon was working with John who was incredibly smart. He had thousands of friends throughout the Americas even if they were not so religious. Hundreds in places like Brazil, Australia, and Europe who did not have hostilities.

Militaries and militias begin to move.

***

Titan woke and expected the burning pain which had accompanied him into sleep. His dreams full of fire and hatred. His body and nerves he hoped were getting used to the searing hot pain. His last procedure had lasted into the day with screams that he did not know that he could produce. The injections were made in very soft tender places which had many nerves. He felt dehydrated, while streaming tears tried to balance the vigorous pain. His mind was actually free, lifted for some time to be enlightened while his body had felt like it would die.

The door opened and a beautifully made up Angelina entered. She was wearing very little which consisted of red, stringy-laced lingerie. She circled the table as if she were staring down at her prey, waiting for the right time to eat while savoring the hunt. Titan was angry but was also beaten, with a destroyed will and ego. He did not want to look at her or hear her speak.

"You're amazing… the first to bear such pain with no results. I may have to put Marcus through the same to see if I AM truly his. He says he doesn't think of her anymore, but we will see." Angelina purred while lifting her head in dreadful thought.

"They will come for me… and Marcus should have been with her… they would be happy." Titan said this as the latter made her angry. Venomous words sprouted from her lips.

"*He is mine. Just like how you are mine.*" Angelina jumped onto him and slapped Titan hard in the face, she scratched his cheeks. This calmed her down a little and she whispered into Titan's ear.

"You will not be saved… you are a casualty of war who has killed many people. You are like my dog that I could make into anything. Maybe a slave… Maybe I will have you work in the mines… dangerous, with a chain around your neck." She spoke slowly with bright eyes, and he did not expect her to start kissing him. He tried to find the opportunity to bite her neck but missed while she took it as play. She moved from his face and started to bite and caress him with increased vigor. He squirmed as if the pain and been brought back and he tried to focus on it. She worked quickly and massaged and caressed as if she was in love. He imagined a succubus, the devil women who destroy everything around them… evil, putrid beings, living in exquisite forms, he looked down and could not believe what was happening. Her lips were wrapped around his cock as she was giving him a blowjob and then she said another amazing thing.

"Its ok baby, I have needles which will make it work." She crept down off the table of torture and crossed the room to the toolbox. She fumbled around for a minute picked up a syringe and said, "I believe this is it... let's find out."

"FUCK YOU," Titan bellowed.

She walked up looking at him as if they were a couple, and jabbed the needle into the fleshy base. Titan could not find words as she skipped around the table throwing the needle away. This crazy bitch would probably just think of any vicious words as 'dirty talk'. He was furious as she jumped up onto him and started into her sensual moaning. Kissing, biting, and playing with him. The drug made him rock hard and it felt swollen, he looked down and didn't know it could get that big, as his member would only stand on vertical. She lowered herself onto him slowly at first. She gasped and moaned clenching her teeth, she looked down at him. He stared at her with cold silent anger... and he could not believe that he was being raped.

Luminescence

Luna now knew the full affect which was outside of New Terra. Military presence was everywhere with no real structure. Some factions had taken over and some had already been overthrown. Many were tired of a crumbling system and had prepared for the quiet strange anarchy. People did not necessarily fight- they communicated and grew. John's people could function almost as if nothing had happened, because they had prepared for the coming apocalypse. Luna could finally see how the destruction of weapons actually created armies and war. Thousands of people were training in the hot desert. This actually made her feel safe from the rest of the world's savages. It amazed her that in this time and advanced knowledge that people would still be so inclined to follow a dictator blindly, still so willing to die for a religious belief, still like a serpent eating its own tail. This enlightenment was not hard to take after Luna's own bout with sanity. She knew that many people would need to be silenced in order to bring- order. Luna felt like she needed to act, and was becoming more worried about her brother, but John did not want to make the first move. He did not want to be the warmonger, but instead would fight for peace and prosperity. John was working vigorously to prepare for the coming fight and he knew that Luna was restless.

***

John found Alluana on the southern border looking into the night sky. She was very still and John touched her shoulder to see if she was awake. She did not turn but spoke to him softly through her angst.

"I had such high hopes for the world... I did not think it was such a terrible place."

John was quick to reply, *"The world is a great place with many terrible people. You know this Alluana. Your country is an example of- good people."*

"It doesn't matter if people can't get along. I don't want to feel this pain anymore, John."

"Pain is just what you make of it... many of the best souls feel this pain. Martin Luther King, Gandhi, and especially Jesus, had taken extreme sacrifice for what they knew was right."

Luna took a deep breath and looked towards the bright, southern stars. She took a moment and looked John in the eyes.

"Will you walk with me through the desert, John?"

His giant head rolled back and smiled as he looked at the stars.

"Of course I will, and may our fate be as one."

They began their journey walking throughout the night. At first they did not speak much, but after a while Alluana began to speak of her own dogma. Nobody really pushed any kind of religion on the Terranian people, but they did make very clear- the difference between good and evil. History classes made moral standards very clear. The present and future would only allow good people to thrive and it is what she expected throughout her life. She did believe in some kind of higher power, but now knew that its effect is only seen by those who want it. Technology is one thing, and can be construed as witchcraft, but she might even believe in God's will? John did not answer but instead choose to nod in silent agreement on everything that she said. Her mind diverted from belief and she began to speak of the things which made her country so unified. How ninety-five percent of the people would go to the Acropolis and night clubs. How all the children would play and laugh with bright futures to look forward to. Everybody was seen as an equal because there was always something which could be perfected or built upon. The cities would pulse with life as people did their contributions which would somehow help the entire world. The few who might become upset would soon realize that it was not a good idea.

One man had hit his children, and instead of jailing him they gave him a choice- to leave and never come back, or to walk around the country twice in the burning sun with little water. It took him a week, and he never touched another bottle, or hit his kids again. This made John smile and he began talking more in depth about his own country, and how it was run. How they actually were in old Arizona which had been a black-ops war-zone for the last couple of years. Many people had been tortured for reasons which they did not understand. The only reason he chose to create a growing empire was to stop the gangland warlords. His people were used to being self-sufficient which he proudly compared to New Terra. They were, however, not nearly as open-minded. He would virtually have to re-baptize them in order to create better moral standards. They would follow him to his death which did not make him necessarily happy. He wanted to be alone from the gaping eyes of his followers, and his solitude had brought what he called sadness, because he could not relate to anybody. His large stature and booming voice only brought worship, and what he truly wanted was a friend- anyone that he could talk to, without being bowed to or gawked at. For awhile he had gotten some pets. A couple of ferrets would become his greatest companions, but all good things must end. Large tears started to roll down John's cheeks and when he looked at Luna, he realized that she must have been crying for some time. He apologized for making her cry which she dismissed, and told him to keep talking. He began again by speaking of the small children- generations of polygamists had caused inbreeding and deformities.

He had to forcibly separate some of the families which had been brainwashed by their own religions. John would take many hours with some of these outcast families, and things changed. Some of the parents would commit suicide and some had killed their own families. This he could not deal with and soon ordered many families to separate and become children of the living god, and after a couple of years the backlash died out, and things were becoming prosperous.

Good things did start to happen after so much turmoil and then the bomb hit. This bomb Luna realized was the destruction of the world's infrastructure... and the weaponry. John spoke of how his country had to make militias to ward off the depression and violence. They had been more prepared for a catastrophe than their neighbors, which is what made his borders grow so easily. He nearly took over Las Vegas, but decided it was too much to deal with, while the gangs made their own turf wars. Then his focus shifted to Phoenix as many reports of torture and murder began to surface. He imagined someone who could be as powerful as himself, but in a very bad way. Whoever it was had eventually contacted him through a tortured soul. The man was bleary-eyed and dazed as if he had run away from fire. Within his ramblings a very powerful warlord was omnipresent in the south, and no one could escape. The man was scared literally to death, as he took his own life shortly after his tale. The cool slow walk had brought insight in the darkness and both were lost in deep depressing thoughts. They were both good souls who were most likely walking towards their doom. Maybe, Luna thought, just maybe their martyrdom would finally bring peace to a war-filled world. This made her think of her history classes and soon she was laughing hysterically in the night. She turned to John and said.

"Maybe a big smile can hold back tears."

His slow, large smile crossed his lips and both were laughing as twilight started to break through the darkness.

***

The Terranian presidents were processing their weekly meeting. Marie Corba was staring intently at Lisa Vanguard and Mark Colburn. Her face was red, the veins on her forehead looked as if they were going to explode. Her fists were clenched with white capped knuckles while she was shaking in her fury. Words finally formed in her mind and she began talking in a shouting tone.

"I CANNOT BELIEVE WHAT I'M HEARING... ARE YOU BOTH INSANE?"

Lisa looked like a deer caught in headlights but tried to be as calm as she could be while replying, "We just want to be prepared. It will be the individuals' choice, of course."

Mark tried to help calm down Marie.
"Yes… nobody is going to make anybody do anything. We just want to give people a choice."

It worked and Marie spoke in a softer tone.
"Ok, only if by choice and only if it actually works. You both know how archaic the world can be and if anything goes wrong then… then it will be scrubbed.

"It is only for the good of our people, and for our own salvation in case…" Lisa spoke as Marie got up and left the room.

"Well, I think it's a good thing to have different views, because she is right, only if it works."

"It will Mark, I hope it will…"

**

Jonah was busy in the heart of Securities maintenance facility when Marie approached slowly from behind. He was breathing deeply and cursing while trying to do some work in a very tight enclosed space. She took her time to look at the massive orb, while she waited for him to emerge from the confined space. She thought that it was strange that he was the only one in the building, but it was late, and it looked like much had taken place. Ten minutes had passed and Jonah started throwing tools out of the strange-looking engine. He huffed and puffed in the confined space and pulled the object loose, but raised and smacked his head on a metal edge. In a minute a bloody Jonah emerged moaning in pain. He cradled the object like it was a baby as he stepped down and placed it onto a table which Marie was standing by. It was a large domed-shaped looking eye with many intricate gears on its base. He was staring at it while Marie was still amazed that he did not notice her.

Instead of speaking she decided to make her presence known by going to the medic station by the door. She returned and was even more amazed that he still did not notice. She wondered if he was like his uncle in a way but pushed the thought aside as she spoke,

"Even the best engineers need to rest and recuperate."

He still did not answer, and she noticed that something was stuck in his ear so she tapped on the table which finally notified him. He groggily looked at her with large raccoon eyes and he smiled through drying blood on his face. She was shocked as he pulled out his ear plugs and started speaking quickly.

"All we need is two more of these, and we will have a militarized IR laser... two more globes, with a hundred planes, and a hundred million volts of energy for good measure. Fifty million amps per shot for three seconds and destruction like no other."

She wondered how such a smart man could become so delirious while she called the base's medics. He kept rambling about numbers and strange things while she waited. They showed up five minutes later… but amazingly they did not take him to any other facility. They shot him up with some kind of drug as they stitched and bandaged his wound. They told Marie that he would be fine as they left. His eyes became clearer and he spoke in a dull, hollow tone.

"I guess I should thank you, even though I get more done that way. It's a drug which opens places in the mind, but the only drawback is that you need an antidote. And if you don't get an antidote in about a week then your brain will bleed and you will die."

"You should care for yourself more than that Jonah, this country needs you."

"Ah, this country needs things done which is what I am going to do. And I know why you're here."

"You… how?"

"Because, I know everything about this place… other than the people, sometimes it's good and sometimes it's bad."

"What do you mean?"

"Awe, never mind me, still lost in thought. The fact is, Marie… that people are conspiring against us, and the faster we move, the better off the world may be."

"What do you think might happen?"

"Well… can't be certain but for one thing."

"Which is?"

"That people are excellent killers and that we are the number one target. Eventually someone will find a way, and we cannot let that happen."

"So you agree with Lisa and Mark?"

"Of course I do. The world isn't dead, arrogant sure, but not dead, and most of this arrogance is based in politics- no offense."

"It's alright, none taken; I just needed a different point of view." Marie's head bowed in defeat as she accepted the unknowing future.

"Hey, it's alright, don't be too upset because I'm not going anywhere. Hey, I finally feel hungry, how about some food?"

They looked at each other and smiled, and took their time to get some decent food.

***

It took a couple of days to cross the vast Arizona desert. Alluana and John now had a growing company of followers who stayed well behind like a shadow. Each day they would wake to find food and water waiting on blankets, which John's followers had carefully placed. They would take their time, and would try to eat most of it, before their long slow journey into the withering heat. They took water but the rest of the food they would leave behind for another day. Luna took a look behind at the peak of day, and she realized their shadow was more like a looming wall of people. She worried more for their safety than her own, and she hoped that her country would not intervene with its incredible weaponry.

She told this to John, and he seemed to not care. They started to pass Phoenix when the first sign of aggression was seen. Gangs were lining the outskirts of suburbia as they kept walking. Within ten minutes faces could nearly be seen as they started to diminish and hide within their homes. John explained that they knew who he was and that they now probably had seen his growing army. When Luna looked back the moving wall had gotten slightly closer and seemed somehow more bulky in the rising heat. They moved along the outskirts of the city and crossed the empty silent highway, and Luna realized that the next few days would bring answers to the outspread question.

***

Titan felt worse than he had ever felt in his life. He would rather have died than been violated like he was. He felt wasted and in a way already dead. He had no ambition to fight when they came and bound him. They took the time to beat some pain into him. More of the Angels gifts which didn't feel like much comparably. They made him walk through his stumbling torture of kicks and punches. It didn't take long to realize where he was going because the den was right outside of the building. Marcus was standing on the edge of the pit waiting for Titan's arrival. This is where he realized they disposed of their victims- in a den with lions. He realized why they wanted him to bleed as Marcus hopped off the ledge and with craze so unfamiliar and maniacal, and he hit Titan hard square in his face. Titan dropped.

"Good times Titan, what a good friend you are. You're going to let me witness true natural beauty."

He grabbed Titan's face and pointed its swollen eyes towards the massive panther.

"Look, it's your old friend... I believe her name is Darkness. And that's what your next dream will be of... you ready brother? Oh wait... I almost forgot your little knife. A friend always gives another friend a chance."

Marcus sliced Titans bonds allowed him to stand and put a small knife in the pocket of the shorts they clothed him in. And in a motion too fast for Titan's tired gaze, he grabbed Titan's legs and threw him over the ledge. His descent knocked him unconscious in the rushing fall that came so quickly. The darkness was welcoming.

***

Samantha was always worried, and in a rage. Titan was the only man she had been with in years, and she was even becoming a close friend to his sister. The only person to help her was the last person who had spoken with Titan. William was not very helpful as he tried to explain where Titan and Luna had gone. Modified deep scanners brought images of an underground passageway with many diverging tunnels. Thousands of paths could have been taken, and Samantha realized that some tunnels were too deep, even for the newer scanner. Luna had been contacted and was surprisingly safe and well. Security finally said that the search was over for Titan. She knew what it meant. He was most likely dead. This did not stop her resolve or her unending search in her own right. She was ranked- so it was not long before she had a meeting with Marie Corba, the International Affairs President. When they locked eyes they were both feeling anger. Marie was the first to speak.

"I wish I could deal with only one person at a time… I know what you will ask."

It was the first time that Samantha began to cry in public. "Please I … I just need to know."

"You're a very strong woman. We need you Samantha. We need you to be strong."

"I am sorry, you are right, but I would at least like to help Luna."

"That's what I would like to hear. We are watching this uprising from hangar 38. If you are able, then go… and do what you can."

"Thank you Miss Corba, thank you." Sam turned to leave.

"And Samantha… sometimes it's best to forget, to look forward."

Samantha dropped her head, shook it back and forth, and left to her destination.

Marie did not know if she should have said that. Her strong emotions gave way and she began to sob. She waved her hands through her holo-screens and screamed. When her secretary called to her, she immediately changed her composure, and was quiet-strong.

***

Titan woke and for a couple of minutes he wondered if he was blinded or if the panther had taken its fill. He moved and tried to feel his torn body which amazingly felt- better. He felt his face and realized his eyes were nearly swollen shut. Dull halos were visible when he tried hard to open his eyes and he realized it was probably night. He could still not hear anything so he sat up and pushed himself to the side of the pit for a back rest. Feeling his face he started to scratch and peel dried, caked blood from his puffy skin. He noticed that his muscles felt surprisingly strong after the inferno that he had been put through, and he wondered why the beast had not yet eaten him, or if he was alone. This thought was quelled when the silence was broken through the powerful consumption of air. The panther sniffing at his feet and the hot breath moved up to his face as Titan became paralyzed with something that was not quite fear. This feeling left a mark on his very small and insignificant soul as he accepted his end to a life well spent. A second later a hot rough tongue started to lick and clean his face. It was unexpected and he was calm as he was ready for more pain. His heart beat slowly as he tried to meditate, he could only think of the end... of darkness. The beast snorted, finished licking, Titan wondered if this was just foreplay for what was to come? Did the powerful beast want to taste its prey? Titan figured he would not be tender... not very appetizing for this primal animal. Why wouldn't it finish, was he actually going to survive? Maybe it liked to run down its food for sport and maybe he wasn't playing the game. In any case, Titan was surprised when it let out a deep, melodious purr, and moved to find a spot to lie down. Maybe it knew he was not the threat. It must have heard his screams of agony for the last couple of days, but he wondered if they were starving the poor beast. Maybe this cat was not so terrible after all, and maybe he should not think of it as a beast. They were in the same position, and maybe their fate would also be the same. The cool night air felt good on his hot body, while he cherished each moment so close to death. Each beat of his heart felt strong and each slow breath brought reconcilable clarity. His mind was free to drift on unexplored planes while he felt dreamily numb. Small was this feeling, worthless and weak, in the encompassing scheme of things. Generations of knowledge had been gained and destroyed in seconds. Hatred, desire, and fate were expunged as a foggy void of nothingness gripped his mind. His emotions faded as his soul seemed to grow into the very plane of this nothingness... nothing was to be held in this world as every falling piece of sand captured another shred of light falling through loose fingers.

Dull light penetrated swollen eyes, and Titan realized that it was daytime, and he could see the black shadow which was his friend. Vibration pierced Titans body as his friend let out a terrific growl. Titan heard the sounds of an opening gate, and witnessed a pacing powerful cat. It fell after being hit with a tranquilizer dart. She let out one more roll of thunder before falling in agony. Titan's body moved without thought. He leapt into motion and pulled the small knife out of his pocket as people entered the pit. Its blade flashed open as the gunman scrambled for another dart. Screams were heard as he grabbed the closest person which he spun around to place the knife at the neck. The others were in shock as the fumbling man could not reload his weapon.

This weapon was dropped by the man who then started pleading to Titan, and Titan realized the person he was holding was a woman and he eased his grip from her neck. Her voice fell out in a rush of words.

"We mean no harm, we want to help, and we do not like that witch called Angel."

Titan realized that these might not be the same people who tortured him. He slowly walked over to the gun and kicked it behind him before he queried.

"How can I trust you?"

Heat

John rose slowly in the cool morning, Earth's shadow moving gingerly across the morning sky. The dark blue was receding in a slowly expanding red and yellow light. Luna was already up parting out the food into appropriate portions. Many fruits and vegetables were accompanied by bread and cheeses. This was a meal fit for royalty in a starving world. Most of it could not be eaten by two people, much less three or four, but it was greatly accepted. The meal was treated as such, and was eaten slowly- and gratefully. She took time to view her surroundings. How her skin had taken on a dark hue while the sand around her seemed pink and bleached in the growing light. Small plants and animals seemed to move and hide from the oncoming heat. Or maybe her eyes were playing with her. The small birds chirping also seemed to play with her ears as she could not compare one sound with the following. She knew that it would be hot out today and did not want to keep walking but for the unseen forces which moved her. A deep calling to help her brother, who she felt was in dire need.

She had nightmares still floating on her mind, a faction of bloody demons, dark things who captured her brother within her dreams. They were grabbing Titan out of the dark spiraling quicksand, screaming-tearing his flesh away. Luna expunged these thoughts as she began to stretch her weary muscles. She noticed that John had barely moved from his position in the sand. He was taking his time while rubbing his swollen legs. Luna realized this and pulled up his white robe to see his puffy extremities. Shock gripped her heart gazing at the massive knees which looked like two basketballs. It looked like they were halfway buried in the sand which must have been dug out for his comfort. She began to cry before she said anything.

"Are you okay, John?"

"Yes… do not worry I am well. Soon my pain will be no more."

This made her sob even more while tears flowed like rivers from her dry body. She curled up into a ball and knew that all was lost without this great man. Thoughts of pain swept through her as she knew that she was the one who made this happen. Her face created pools of wetness held in her hands. She heard movement close to her- John stood.

"I have taken pain through my entire life. Sometimes it allows me to feel more human."

Luna looked up at the towering statue of a man. She noticed his glowing eyes which made hot tears dry as she rose to meet his gaze.

"I have cherished my time with you and am glad that I have taken this journey. It is not your fault that mine will end soon."

She wiped her eyes and smiled before speaking in her soft voice, "You have restored me, if only I can be with you always."

"You will always be in my heart Alluana... and now, will you walk with me?"

She grabbed his massive hand and grabbed onto his waist to try and help his stride. A mumbled gasp echoed from the people behind, as they made their road in the hot blistering sand.

***

Titan calmed down after the man begged him to take his life instead of his daughters. He realized that they needed help, which they would not get from their own countrymen. Within minutes they started their journey to what Titan hoped would be a sanctuary from the evil that he endured. His companion was again a sleeping panther in the back of a small van. The tranquilizer could put down an elephant and was expected to last for four hours. This did not make Titan feel more secure, for if this beast was to wake he would not be as lucky as he seemed to be. Fortunately their destination was only an hour or so away. This gave the woman time to speak of the dictator who was Angelina. She told of how this devil respected gangs even more than the fellowship of the faithful. And all was quiet around the compound because she was going to wage war on a growing army to the north. A sanctified man was challenging Angelina's power. This made the power-hungry woman furious beyond belief which could only mean one thing, bloodshed. They had already felt relief from Angelina as she pulled her mob-like militias together. They prayed to God for the quick destruction of this woman who never brought any good to anybody. Titan wondered how Marcus had fallen into her good graces, but did not say anything about it.

Quiet had passed for some time but before their destination was made the woman asked Titan for help,

"My sister is a slave to the gang called Los Locos. Will you help save her?"

She had turned and said this with tears on her cheeks. Titan's blood boiled as he thought of his own lost sister. He must have looked angry as the woman had turned quickly so he replied in a dry raspy tone.

"Of course I will."

Titan's hot mind raged as he tried to place himself in his meditative trance. The numb rush of nothingness brought calm, but now he could feel his pulse. He realized that his calling was not nearly over. He thought of Marcus's words when he was racing through the city. 'Scum is the fat of the land.' Titan thought these things with fists clenched into white-hot knuckles. 'This scum will soon be scrubbed clean Marcus, my job is not nearly over,' He had to meditate to stay calm as they entered a small village on the side of a jungle. He told them they should let this panther loose unless they want to be like the demons they hate. The old man did not look pleased, but after-thought said that Titan was right. They drove through the town and into the forest for twenty minutes until their trail stopped. Titan picked up the heavy beast with ease and placed the mighty panther on a soft cushion of grass in a hidden ditch. He had never been so close to such primordial power but felt as though he was leaving behind a newly found brother. He caressed the panthers head and mane before he left, and gave the beast a kiss on the nose. He could relate to this shadow of death, because they were the same, branded by the same nightmare. The man looked at him in shock, and then with something that he could not quite place as they made their way back to the small town.

Chapter Three

Red sand

The desert was hot with scorching waves rising from its reflective surface. The midday sun felt close when Luna rocked her head back to glance at the ball of fire. Her body felt a shock of cool sweats. Her mind swayed and she placed a knee on the sand. Cramps locked onto her hands as her mind rang in a rolling twang. She tried to focus to calm her mind and her hands.

"Are you ok, Luna?" John asked as he rubbed her back with his tender touch.

His voice reverberating through her ears as she pushed it aside to find order. Her body was trying to combat the relentless heat, as its cool sweats tried to relieve her desperate feeling of anxiety. The mind trying to push through its ordeal as her body felt a mellowing relief, and wiping sweat from her brow as she regained composure. She looked at John who seemed fine as hot waves rolled around her eyes. She could not tell if it was heat or if she was out of focus. She stood and remembered to reply.

"I'm fine, just getting used to it. Don't you feel hot?"

"Yes, but I'm fine... should cool down soon in a few more hours."

Slowly they walked up the ridge which Luna thought would never end. Stumbling into the soft moving grains of sand, this shifting earth would not allow for any substantial forward movement. Luna wanted to crawl on four limbs to get it over with, but relaxed when she looked at the calm stride next to her. Eventually the crest began to break into the view of the new horizon. Luna quietly rejoiced as she thought she saw small trees in the distance. Trees or cactus but anything with shade would be accepted. The distant shadows waved dark like a wall of moving ghosts. They rested from the torturous travel as John stood tall to look in the distance. Luna finally looked behind her into an oasis full of people. Like an exodus from ancient days she imagined, a repressed people who will create future gains from power in numbers. She smiled as she turned to look into the dancing cacti. John's head was solemn and bowed as he looked into Luna's brightly lit eyes. Her smile was quickly reversed as she realized her questions were answered before she asked it. Her gaze tried to focus on the moving cacti as she realized it was a large group of people. Chills crawling down a perspiring back and she knew this was the Angel, the thorn of power which they were irritating. They were the only people to rise against this unknown threat, to make a showdown for power, in the sweltering midday sun.

***

Titan was restless in his new surroundings. The place was dirty and repressed as he looked around and spoke with his new friend. Children laughed and played in the small street of the village, while women were busy cooking and cleaning. He tried to focus on the map laid out in front of him. Guns and ammo were stocked in the well-built shed behind them as Luis spoke of his son. Soon they would be back to help and tomorrow they would save his daughter. Titan did not want to wait as his freedom was most welcoming. His body wanted to act and his mind wanted to help whoever he could. Mentally he had to meditate as he stifled his imagination of a bound and tortured woman. This woman was most likely being raped in mind and body, and was probably already lost for a better life in her captivity. Titan's fury was cooled by hot soup and warm smiles, and as he and Luis went over plans and maps Titan began to create his own in his mind. He knew that nobody would expect an early morning raid, much less a rescue into a warlord's home with only one person. Titan realized he may not survive but a one man rescue might actually be easier. 'A silent attack might not even be noticed at all.' They went back into the shed to go over the armament that they would use- rifles with scopes for long distance sniping, small machine guns for quick death and takeover, and a large assortment of semi-automatic pistols- much easier to carry. Then it was spotted by Titan like a shining star- the hilt of a sword suddenly stood out of the darkness. As he went to pick it up, the old man smiled and laughed as he unsheathed the dull slightly reflective blade.

Luis laughed, "You will die if you use that against guns."

The old man explained that they had gotten it from a Pacific Islander who sought refuge from his eastern homeland. Titan realized it was professionally crafted as the old man rambled on about past adventures. Titan put back the sword and told the man that he would need some rest for the coming day. The man smiled and handed him a black pistol, he had already been shown his new home in the small back yard which was a tent and hammock. A perfect get away- especially if he could move a motorbike from the street silently and wire it. The midday heat was cooling as Titan lay down to rest.

***

The number on the screen displayed close to thirty

thousand. Samantha was trying to locate Titan through the satellite's imagery. Luna's escort looked at first like Titan, but Samantha was actually relieved when it wasn't. The Mexican army had tanks and mortars hidden in the background waiting. Samantha knew that it would be a massacre and was hoping that Marie would soon take her call. She sent out a constant live feed of the two armies, or more accurately- one army waiting for blood, while the other had no more than a few thousand guns more suitable for a gang-fight. A call came through as Samantha witnessed a small group of people walk forward to meet Luna and her unknown escort. She had to focus on the call.

"Yes, this is Sam."

"This is Mark. Marie is busy Samantha, but I am watching your feed now."
"Please, Mark, let me help Alluana and these people." Sam tried not to sound like she was begging.

"You know I want to, but it has been hard to convince the panel. They say we need to stay out of the affairs of others." Mark said in a dull robotic tone.

"That's as bad as genocide, are you kidding me? Luna is a citizen and she needs help, these people need to be saved!"

"Any war could be genocide… hold on, I have an incoming call." Mark said this as he put Sam on hold. She felt her face go red as she could barely hold her emotions. Was this President just messing with her? She noticed the people falling back as a large crowd of twenty thousand came running up to figures lying on the ground. She missed it, and Luna was dead before she could do anything. Tears sprang from her eyes as the two people looked red in the hot haze of the day. She knew the real slaughter was going to come when she got a voice.

"Hello?" It was different man.

"Hello, this is Sam, I need green codes."

"This is Jonah who has taken over Security with General Hollins and Admiral Nash. The judges and their council cannot command. You have green."

"Thank you, sir." Sam could not believe her ears as the codec pulsed green and she focused on her targeting system. The comfortable chair was attached to a visual marvel. This targeting system was only unlocked by a series of codes which could not be overridden. This was a room that controlled incredible power which even scared the power-in-charge. It was a strange sensation to manage such power which the entire world would envy. Sam did not think of these things as she backed up her view on the moving trucks. The creeping mobile weaponry with heavy machine-guns, for a quick and deadly extermination. The targeting info was calculating terrain for the least amount of casualties while the weapon craft took position. Sam knew that falling debris could be just as deadly so she tried to tune the position as she pressed the button. Its lag disturbed her but soon felt a shock roll through her veins as the visual hit her eyes. A bubble evaporated a massive gathering of artillery and people, the rest fell for cover from the blast. Its dust blinded her from normal view but she could see small movements from heat signatures as people who were in shock and awe. She knew she couldn't do any more as it could only do more harm. She hoped it was for a good cause. The next call was also unexpected.

"This is Admiral Nash, is this Samantha?"

"It is sir."

"That was a good effective shot which was justified, commander."

"Thank you, sir."

"Even Security has to change, because we cannot always be run by people who are in the dark. You will report to me while the transition takes place, ok Samantha?"

"Yes sir, and please call me Sam."

"Haha, ok, I think…" Static hit the communication headset which Sam thought was weird. Did I make the Admiral angry? Was that not really a laugh? Why would he cut out like that on a secure line? She felt worried but that feeling was soon quelled by the image on her screen. It was focusing on what looked like a small planet in the sky.

***

John and Luna had been quiet for some time staring into a certain death. Both were comparing odds and both realized that they were going to be martyrs for a greater cause. Luna wondered if the politics of her country would overlook their demise. She figured that if Titan was home then things might have been different. And in a way she had only helped in the incoming destruction of the people. She wondered if these people had Titan and if they could make a deal. John did not speak until the entourage started to approach. His voice was calm and clear in its deep roll.

"It has been a good journey, Alluana. And I wish that things would never end, but in a way for me, it will be welcoming."

Luna could not believe her ears. She had wished for the same thing in the last couple of months, but now she would rather live.

"Please John, don't speak like that. We will negotiate and keep living."

"You have not heard of the stories of this crazy warmonger. Some do not deserve to live."

"And some do not deserve an early death, and I would rather die then live in slavery or injustice."

"That's exactly how you will live your life. Free and with no regrets. Don't stop for anything Luna."

"I won't stop. I'm going to save my brother."

"Yes, that's exactly what is going to happen."

"I love you John, you are the best person I know, but we are lost."

"Do not think like that, God moves people for reasons that we should not question."

John smiled in a wide grin with white flashing teeth. They both smiled as the two trucks stopped to greet the large crowd of visitors. Luna's smile quickly turned into a frown as she saw a crazy bearded Marcus step out of the truck after a woman. The woman walked quickly- to speak first in angry glory.

"This is my land… **How dare you all. You will all perish like dogs**."

"This land is not for one person." John's voice rolled back.

"The only reason that I come out here is to see this rebellion crushed like the maggots that you are." Angelina was looking back and forth at the people in front of her. Luna wanted to try and negotiate with this woman and especially Marcus.

"Are you Angel?"

"You can call me Queen, you idiot!"

"Please, my brother is lost. I believe that you might have him?"

"Ah, this gets better… Marcus, my love, do you know this whore?"

"Yes, Angelina, she has a brother who will serve a meal for the beast." Marcus laughed in a hysterical tone and stopped. Luna looked broken and tired at this news.

"What have you done to him, Marcus? Please…"

"Your brother has returned to the circle of life, Luna… He will be energy for a large panther named Darkness." His grin pierced her soul as he said this, and she screamed in anger.

"NO!" Luna lunged with a weakened body at Marcus. He pushed her back easily into the dry sand and laughed.

"Bitch, your brother deserved to die!"

"No, he was doing his duty for this world, and your love did not deserve to die."

"You can't even remember her name."

"I remember everybody's name Marcus. She was sweet Katie."

Marcus drew a large knife which he shook in anger. John stepped between Luna and the blade as the quick agile Marcus moved in a flash. Luna stood and realized that the blade had already been used. Blood gushed out of John's side as the metal was taken out. Luna once again screamed as her voice seemed to echo and intensify. John's people who had moved close realized that John was hurt. The throng began to charge in a screaming frenzy. Marcus was swinging the blade and grabbed for his pistol as he backed up from the growing roar of footsteps and screams. Angelina had seen it coming and both fell back into the entourage.

Luna tried to grab John before he fell. Blood flowed like spilt wine into the spongy sand. Luna held onto his wound but the blood pulsed as his strong heart beat. She could not focus on the wound trembling-she tried to calm and bandage John at the same time. With shaking bloody hands she told John that he would be alright before he said one last thing.

"My time is good, speaking with you as equals."
"I would like to have seen your country Alluana-with towers much larger than I."
"God... Will... Save us."

Alluanas tears were streamers as his followers were nearly upon her she kissed his forehead when the shockwave rocked the ground from under their feet. The weapon crafts rail burst was a balloon of dust and debris as the people slowed down in a shock and awe with the power they had felt. Marcus and Angels escape had vanished into the fog of falling dust as calm was felt before another massive force was made present.

A fresh light burst through the afternoon sky.

A new sun was born through the heavens with a pounding wave of radiation which made the air burn. Wind began churning in a tempest of sandy dust devils. People began screaming hysterically for their savior. Luna was being crushed by a pile of angry mob who wanted to see their lord. She was trying to do CPR in the tightening squeezing air. It was squeezing until her lungs burst into a deafening scream of anger which made the people stop. Quiet fell once more -second's maybe- before a rise in light brilliantly illuminated the area around them. Everything and everyone –glowing- in a bright and luminous blue-red and weeping Luna was heard through the silence of the throng.

Eye of the storm

Titan had tried to make like he was sleeping when the light brightened around him. People were moving and yelling around him in shock. He rose and realized everybody was looking into a new sun. How was this possible he thought? They did not have any more ultra-heavy material. Was it an aftershock? A year after the interstellar anomaly was destroyed? Titan moved behind the group of villagers and realized Luis was in the front of the crowd. It was his time to move while people were busy. The shed was open and he took the map and he took the sword and glancing at all the weaponry- he already had a small pistol and he ran. He moved up the forested hillside as the heat began to rise. Blistering radiation began to hit Titan on his nose and through his throat but he did not slow down. His destination was only a couple of miles away. He decided it would be best to use the forest as his road. Small trails wove through the forested hills while he gauged his distance and direction on the map. A new sun made it hard but his feet would not stop as he cut his way through the brush. Heat boiling through his mind and around his body as the distance was eaten by his newly found nature. In what seemed like no time to him the forest finally ended at a large field, a house, and a town in the distance. 'It couldn't have been more perfect,' Titan thought. The house was more like a mansion with many people standing around the entrance looking into the sky. Titan knew exactly where to go and what to do which felt odd while he made his way to the back door. It was already cracked open and swung inward easily. His infiltration felt too easy as he made his way to the basement where the cellar door would hold captives. He finally slowed his pace as he crept down the stairs and heard voices. Titan hid behind stacked boxes and shelves as he eavesdropped on the people. A man and woman were going through plans over a small table. It sounded as if they were annoyed at Angelina and her constant dictatorship. They were planning blackmail and murder. Titan widened his eyes, squinted, and realized that this woman was the one who he should be saving at the moment- Luis's other daughter- empty chains on the wall. He was going to be used as a pawn in a game for power. The woman was going to be used as bait and for an excuse. Titan was going to be used as a commodity for whatever reason. His boiling anger started to grow as their voices sounded resolute. Images of the powerful panther flowed into Titan's mind as he remembered its unbridled primordial power. 'Maybe he should not kill these people' he thought. 'Maybe he will just maim them.' Then he wondered why they were not informed by the old man Luis. Was this not the right daughter? Or they just haven't realized where Titan would go yet. Motion happened in a flash of steel as Titans muscles could not be kept still. The man's head fell off of the shoulders which made the woman faint as he stepped into view. He was glad that he didn't have to stop her from screaming as he looked around. The basement did not even have a cellar door- but a wall did have empty shackles. He was furious. Why would they even go through the hassle of it all? He went back to the table and rummaged through the plans. Pictures of what looked like a death list was present before Titan threw the table against the wall. He gagged the unconscious woman and roughly threw her over his shoulder. His exit was just as easy through the back of the large mansion, and as he disappeared into the thick brush a van pulled into view of his position. The crowd welcomed Luis, the same man who saved him. Titan was betrayed and they were going to somehow make him into a commodity- some kind of leverage against their dictator that was Angel, a patsy to get close? Nobody would care for him if he was used and the woman he carried must be powerful with a large family. 'Maybe she was in line for the throne, and maybe they needed bait.' Titan questioned these motives as he moved through the brush. This country was controlled by many small gangs but the top was always going to be fought for. 'What idiots' he thought as he made his trail. How strange that the world would revert so easily to a tribal gangland.

'They didn't make very good plans.'

***

Luna screamed orders as the people realized that she was the new boss. Strong men with weapons were told to kill anybody around the crater who may have survived. Men who did not were told to start digging in this very spot. They were told to get lumber and tools as this was their new home. Women were told to set up camp in the desert, and to make food and make sure that the people's wounds were mended. Soon the area cleared as people left home to gather supplies or fight. Angelina's troops were scattered, fled or dazed by the massive shockwave and guns began popping. The slaughter was not fast and screams rang in the distant ridge. Those who already brought supplies started to make a circle of tents around John and Luna. Luna's eyes were swollen as a new sun burned the sky yellow and her tears seemed to dry as fast as they were made. She felt as if the drying blood was creeping around her body as radiation penetrated bronzed limbs. It also felt welcoming as she mourned the large man who still commanded his people. A village grew out of the desolate desert in what felt like no time at all. The light should have already faded when Luna realized how many hours had passed. She felt tired and wondered if she was dreaming in the hazy fog of the heat. This strange light seemed to penetrate everything around her. She lifted her head from John's chest and realized that her people had covered them with a canvas awning which did not shade them from the reflection of the sand. Every direction seemed to glow in the sun as Luna rested her head. Her dry eyes closed as an uncomfortable sleep enveloped Luna's mind and body.

Hours passed before Luna woke to darkness. She wondered if she was finally dead, or if the game still had to be played. She slowly stood through her achy body. Her head hit canvas, and she realized where she was before she saw the massive body. A moment passed as she stared in disbelief at the irony of it all. How death could create such a following of life. She looked down at her red robes which used to be white. She wondered why she had to still be alive. She would rather be in the peaceful realm next to John, walking to a better place. She dissolved this thought as she heard murmurs from around the camp site. People had made a large circle of tents, but many were awake sitting within this large circle. She could not guess how many, but knew that hundreds soon equaled thousands. And as her eyes focused she realized they were looking up towards the sky. A few aching steps and she found herself looking up into the heavenly phenomenon. The new sun was apparent, but seemed as though it was absorbing the light around it. A swirling eddy looked like a large tunneling vortex. As if space was collapsing in on itself while light seemed like it did not want to escape. Chills crept through Luna's spine, and she wondered if this was really Armageddon. Would the world be eaten by devils for all of the people's sins? Would anybody survive? She thought that maybe it would be a good thing, a possible rebirth into a new dimension. A place where there was no bloody wars... peaceful, with no ego for power for objects, a fresh new world, where people would actually try to help each other, to live full happy lives. These thoughts made Luna smile and she screamed as loud as she could,

"OUR LORD HAS DIED, AND SO A NEW LORD SHALL BE BORN AGAIN!"

The people stared at her with fear in their eyes. They whispered and began to pray to themselves when it struck. Fierce winds began to churn the sand as the object in the sky imploded into a halo of light. The still was just as quick as the light emerged to overcome any dark place. Thunder boomed into the atmosphere when the energy struck. Intense gravity slapped the earth and lifted the air from the ground. Luna looked back to see the canvas that had been blown away to reveal the large body. What came next brought fear to all who were now awake with open eyes. John seemed to be the first to rise off of the ground as people realized that the earth no longer held gravity. They were all floating in the sky for a moment before the earth regained its composure. The energy of gravity rebounded in its magnetic force. And all fell back from the few feet of weightless floatation. John, however, looked as though he was sitting while people bowed down on their knees in a weeping prayer for their lord. John slowly lay back down as Luna walked over to kiss him on his forehead as she closed his eyes once more.

Armageddon

Samantha thought it was a missile at first so she hit the trigger. She then realized her telescopic imagery had gone off the charts. This 'missile' was hundreds of thousands of miles away. Her weapon craft shot but she did not know if it would even do anything much less even travel that kind of distance. The spherical light grew into a new blinding sun for a moment while her imagery focused on different spectrums. Samantha's heart pounded as she realized that this was the aftershock of what must have been the disappearing anomaly. Waves seemed to fold out of some unknown place as spectral layers were focused into view. A new sun seemed to grow out of nothing. She got a greatly accepted call.

"This is Admiral Nash, do not fire until we know

what we are dealing with, ok?"

"Yes, sir. What is it?"

"We are contacting techies now, and your guess is as good as mine."

The comm. link was silenced as Samantha stared into the new phenomenon. She could not tell if it was growing or if it was eating up space. She compared the light spectrums with the Suns and realized that it was nearly twice as intense. She wondered if the atmosphere could deal with this and realized that it wouldn't matter if this thing decided to come close. Life would perish if two Suns were combined and chills ran throughout her mind and body. It was beautiful she thought, like a deadly rose blooming out of the sky. Thorns placed behind its lovely exterior which screamed eternal power. She focused and through the blossoming light she saw an empty black void. Like a wormhole- and she realized it was a black hole. Or at least something that may be similar. It seemed like a long time before anybody was in contact. She looked in the dark around and at her copilots who controlled their own craft. Their faces looked white through the dark of the room. They controlled the most powerful technology in the world, but it all could be destroyed in one stroke from the cosmos. Samantha breathed deeply and accepted her own fate. It relaxed her as she got the call.

"This is Jonah. Our meteor is back."

**

Titan moved through the forest with ease, even with the limp body on his shoulder. He decided to take a winding path close to main roads. He could only imagine that they would not even care about this woman who was just as much a pawn in their game. He figured that they were most likely trying to sway gangs to join in their crusade for power. He still could not figure out why it had to be such a game for these people. Must have been some how bred into them. After all, people are more willing to act with anger on their minds. He knew this to be true in his own blood as he stopped to breathe in a small clearing. As he crouched to lay the woman on the ground she moved and grabbed the pistol which was strapped to his leg. He had forgotten it was even there as he held up his hands. One hand still clung to the sheathed sword.

"Drop it," she said with crossed brows.

"Maybe you should just shoot me I have no where to go," Titan replied with a smile.

She humored him and shot. The bang was loud and Titan immediately felt for the wound. There was none and he realized that it must have been full of blanks. She looked at him and realized who he was. He was the man who was going to be their scapegoat. She dropped the gun and started to run out of the clearing and abruptly stopped. Titan stood tall and focused past her gaze. A large shadow moved behind the edge of the clearing and he realized that it was his primal friend- the panther-Darkness. Titan's smile got larger as he heard the low rumble of a bone chilling grow. The woman walked backwards until she backed up into Titan.

"You don't want to meet my friend, the Shadow?" Titan asked as she turned.

"Please, I beg you, please don't kill me… PLEASE." The woman said this as she started to cry. Big burly elephant tears sprang to the ground as the hot sun dried them quickly. Titan's smile turned into a frown as he tried to stifle his humanitarian emotions. He spit on the ground next to the crumpled woman and for his friend. The Shadow cat was nowhere to be seen. Titan looked back at the sad, selfish eyes of the woman and spoke.

"Fine, I need to find my way back home and you're going to help."

"Oh thank you, thank you god," she said as Titan began to walk. He wondered if it was he who she thanked, or whatever god she might have. He asked her where they were going, and now she was more than willing to speak, as she realized that he did not care about her.

"What you don't really know? We are miles into, and are heading further into the forest."

"Belittlement won't work on me, woman. It's either the forest or those people you call human."

She smiled. "Those people are my family and it's not the first time my boyfriend has died before my eyes."

Titan was amazed. He expected that she was no older than twenty. "I didn't kill for power, I killed for a message."

"Yeah... that message made me hot for you, big man," she said this as irritatingly as she could.

"Don't speak unless it's for direction. All I have dealt with in Mexico is pain and headache."

She began to play coy. "Our direction is towards the forest. So killing relieves headaches? Good to know."

Titan spun around as he had had enough. "The only one I can trust is that panther. At least it doesn't play games, unlike your piece of shit family."

Titan began to walk faster as he heard the sounds of whimpering. He stopped and put his chin on his chest before he turned to look at the crying woman. 'Why was she such a pawn he thought?' And quickly knew that his own-self was kind of the same way. For a minute he stopped to think but knew that he should keep moving, so he did. And fast was his pace as he fled through the forest. With people on his trail, and an unknown shadow at his side, he moved.

He stopped so the woman could catch up for just a minute. Then he moved like the wind and the rain, relentless through thick, unknown brush, far from the civilizations of so called humans. He and motions moved beside him while people played chess and pawns for whatever reason they may please. Money and trade loosing power with each step, Titan, like the rest of the world began moving for survival.

The woman was nearly keeping pace, which almost surprised Titan had it not been for the struggle in this land of theirs. And while Titan ran with a maddening pace, he wondered why his veins did not boil as in the past, and he stopped. Everything seemed still in the strange twilight of the night. He heard a shout and moved toward the sound of voices which grew out of the next small, mountain range. Thick, jungle-like forest parted as he moved his blade and body to part brush to peer into the sound of the glowing night.

People sat around a large stone idol, blank faces staring at the sky. Titan wondered how this could be possible in the middle of what seemed like nowhere. And he realized that these people could be like any who wanted to survive- tyranny free. He wondered what the best course of action could be when the woman crept up behind him. She was trying to not breathe hard as she made her way next to Titan. He asked her what her name was. She told him that it was Penelope. He thought it was a beautiful name and he realized that he should say it even if he did not like her at the moment.

"It's a beautiful name." Titan whispered.

He looked at her before he walked towards the people in the middle of the clearing. He then wondered where the night went as he raised his head and saw the hole in the sky. The yellow sky was apparent as he noticed how the forest let the light in and to the ground. The light seemed to be rising. Titan's gaze was on the light which seemed to be swirling in the dark void of what should be night. He realized that this was the new sun that had been born which made his journey so sweltering. Growing light that made him want to look up at its radiant intensity in the still air. Air that seemed pure and crisp and air that also made him realize that something was going on. The woman named Penelope looked worried as he smelled the clean ozone smell. And an old man on patrol around the camp finally noticed the two people standing by the outskirts of their village. A loud horn rang as Titan and Penelope crouched from the unwelcome sound. Arrows fell around Titan's head and he realized he needed to moved but couldn't so he screamed a reply,

"STOP... WE MEAN NO HARM!"

The arrows stopped while people ran to get more weaponry. Soon they were surrounded by people who looked like hermits. Arrows and guns pointed at their faces as Titan dropped his sword, and both raised their arms in the display of capture, all while a brilliant light began to strangely grow around them.

"WHO ARE YOU?" The old man asked who first spotted them. Titan lowered his hands and spoke.

"We are both refugees who escaped torture and murder."

The people eased up as the old man asked,

"What is going on around us?"

"We don't know, but are glad that we are alive to see it."

The people seemed to drop their weapons as they realized that these intruders did not have any upper hand. Titan, Penelope, and the people looked toward the sky into gleaming light which seemed to be swirling like a torrent in the sky. Thunder penetrated the heavens like a god's terrible growl. This was followed by energy which seemed to pull them from out of nowhere as the group started to feel weightlessness. Titan thought that it was similar to what he had felt before as air seemed to lift from the ground while his body felt buoyant. Unbalanced force carried them onto an unseen blanket as they shouted for answers as they rose. A diminishing radiant light fell on all of the astonished faces as it made everything glow. They had risen and fell together and knew that the world would never again be the same as it was before.

Chapter Four

Adam Jones was no longer with the world, but his extreme mathematical equations would be used for generations. Super computers were created to be able to crunch millions of overlapping algorithms. Nano-molecular engineering made way for small, powerful processing units and engines that would be used in virtually everything. New Terra was even very open with its technology while other countries could not duplicate the processes that capitalized on the future. Now these lightning fast computers were growing hot as teraflop's were computing the massive amounts of energy coming out of the fabric of 'nothing'. The swirling eddy was trying to gain matter- it wanted to live. The weapon crafts were then ordered to fire at this thing so that it could gain the power that it needed to move out of the twisting vortex of space. The small pellets of volatile material were shot soon after the supercomputers decided that it should be all or nothing. Tens of thousands of projectiles were shot at this body of power while people waited for the light to dim. Within a day the material met its match and energized the vortex of the dark anomaly. Its power reached a critical point while it decided what to do with itself. Most light was unable to escape the swirling event horizon, while its heavy material decided to reform into reality. An explosion of energy leapt out of the vortex as two beams of power pushed itself in opposite directions. A wake of gravity pulled from behind these energies. It was about as far as Mars while fortunately the Moon was on the opposite side of the earth. Many thousands of earth's gravities would not even compare with this extraordinary power. And fortunately the earth would not feel most of this unforgiving power as the two asteroids rocketed out of the solar system. The earth would survive but the damage was all ready done. Millions of tons of water had been carried on the blanket of energy. It fell back only to meet intense and magnifying waves.

Three atoms of death

It could have been much worse. If the moon had been on the wrong side of the world, the tides would have been even more devastating. Or more gravitationally effective as it would have compiled energies, but fortunately the moon pulled on the tides from the opposite side of the world. Many had felt a squeezing sensation from the opposite affect of weightlessness and pull. Within hours massive waves had combined into thick walls of mega tsunamis. Many were as high as a hundred feet or more. Wave sensors had been going off as soon as the gravity had pulled on the oceans. Tsunami warning sirens were blaring in virtually every beach in the world. Many people would not take notice, would not move fast enough, and many stayed behind to stay in the homes that they loved in deadly ignorance. People would ignore things just because they could. Most where stuck in the middle as people fled for higher ground. And many places and people did not see the terrifying effects that would be the deadly ocean.

The swell was fast as water receded from the beaches to leave a ghostly calm before the wall of water was seen in the distance. Those who stayed to watch knew that this would be the end. Their lives flashing before their eyes as the ocean crept onto land before the liquid titan devoured everything in its path. Hundreds of thousands of lives were lost in mere moments, while the rest of the world braced for their own deadly impact. The Pacific Ocean had felt it first while people on the Atlantic and Indian coasts thought that they had been fortunate enough to escape. They only had more time while people were arrogant with their fragile lives. Most could not imagine such a destructive force. A force that made all life possible, but its power was devastating in its unforgiving form of irony. By the end of the day the death toll around the world had been taken into the millions. Those who were fortunate had land on their side, or if lucky, strong flotation devices to counter the powerful torrents of water. Some who had a powerful will would bear through the relentless taker of life, only to witness massive graves on the beaches of those they loved so dearly throughout their lives.

Days had passed as the water receded back to an eerie calm, lucid glass, as if the ocean felt humble and tranquil after its devastating attack, leaving behind lost souls devoured by a liquid nightmare.

***

Alluana and her newly found peoples were fortunate to not feel the effects of the devastating oceans. They felt pain in their own way, while they built a temple around the grave of their fallen savior. This grave was the first thing to be built as thousands more had gathered to mourn a great man. Large oaken boards had been cut and were intricately carved to bear the greatness of the man inside. And while he was placed to rest, the rest of the world had braced for an unnatural catastrophe.

Work did not stop with the burial as Luna and her peoples began to erect the skeletons of buildings in the hot oasis. Days passed and an entire city was formed from the growing masses of people who were looking for answers, in a most volatile and depressing time. Word spread of the anger that God had sent on the land for the death of God's beloved son, bringing death that would not be overlooked by the many rising faithful. Luna and many of her people were making plans to rid the world of its terrible demons.

Overlords who gained power from the pain and suffering of others were now the ones who crept into hiding. They hid in the same places that kept them safe for generations from the might of unseen forces. While they had crushed and conquered the individual will out of the people who helped them. The people's will was now renewed by many who no longer felt the umbrella of evil. And while a new dawn unfolded a massive gathering of peoples encompassing the miles of sand outside of the perimeter of John the Prophet- the Savior.

Luna welcomed the new peoples as she walked around the newly built city. Some she noticed were even in clothes reminiscent of her past country.

***

Samantha was like anybody, shock and awe brought deathly chills throughout the dark control room, full of people, who were looking intently at the disappearing beams of energy. They helped save the earth for now but the light of day was most unwelcome as they braced for the rising ocean. Red sirens blared throughout New Terra as many who were underground sealed themselves into the underground maze. Many were being taken away by the fast air transportation as many jets were busy picking up those who would want to flee to higher grounds. As dawn rose, so did the water on the flat plateau that was the land of New Terra. Samantha and a few others decided to leave as the small jets landed on the roof of hangar thirty-one. Jonah hugged her and placed some wrac keys in Samantha's before she realized that he was not going to leave. She looked at the group of people and knew that they would rather die then leave the beautiful country that was their home. She wondered how many would stay behind to accept whatever fate may lie in the swamped land. She wondered if the underground city could withstand such pressure from the deadly rise in water. And tears flowed as she left the only place that she ever called home, the place that would always reside in the deepest places of her heart and her mind. And within an hour her flight let up from the intense pace of its speed. A great, green pasture was presented before her as her new home grew out of the dense forest of Brazil. A small peaceful, quiet village, with kind generous people who would take new visitors with open arms and hopefully quiet mouths.

Jonah and his team of high ranking security officers had already taken an oath for their country: To defend their home against all odds and anything that would threaten their existence. Now this threat was again a natural force which screams deadly power. They had traveled into the bowels of their city, as water rushed around as they sealed themselves into the only place that they cared to be. Large pumps were turned on as they piloted the weapon craft into their swamped refueling cradles, which could easily be their final resting place. Soon the last had entered as a shower of water allowed for its entry while the water created plumes of steam from its hot engines. The gleam of deadly power was reflected as Jonah powered down the last weapon. Devastating weaponry, that was able to destroy the majority of the world's power, and influence in a matter if days. For now, Jonah and his crew were doing what they could to command from underground, so that people may survive the deadly calamity. The builders of New Terra never would expect such a rise in sea level. But they did engineer the buildings as well as they could, with an underground skeleton which had to pump out ground water anyway. And what really saved lives was the height of these towers where most people lived off of the ground. Jonah knew that he and his crew were the ones to be in real danger, but he also worried that now they were vulnerable with no defending weaponry. He and the generals had planned on things of this nature, but never expected such an unbecoming force. It was very detrimental to the security which was a top priority and they were wanted people. It didn't take long for the military to completely take over everything. Even most of Jonahs clearances and by the end of the night most began to treat him like a child. Anger mounted as he recalled all of the many hours of hard work in the different military hangers. Technically he was not military, he was a contractor but he had roots from the foundation of Terra's beginning. He even considered himself a founding father as he had known many of them- including his own grandfather. Now to be pushed away after everything seemed to be falling apart. He suppressed his anger and found himself waking to his own bed in his apartment. It had been months since he had been in his own home. So quiet he had to at least open the blinds which let in a flood of light. It was warm and made him smile and he knew he would never leave his only home.

New beginnings

Titan had finally found a sanctuary from the evils that he had endured. It was a small group of people with a little over two thousand strong. Their will to live would outshine their repressed looks. They lived very well in a dark, dense jungle, away from the politics of the outside world. Titan was accepted with open arms as they learned of his glorified homeland. All peoples had heard of the incredible technologies which grew out from a brilliant city. But now their focus was put on the woman named Penelope. They did not trust any outsiders, especially those who were from the known ganglands. Many wanted to kill her on the spot. Titan's strong opposition was helping her survive. They wanted to shackle her in a large pit which was their jail. Titan would not have this at all, but only kept her from being bound in the small jail. And soon they would allow for her release as long as they felt like they could trust her. Titan now realized that he would have to help her get out of a place that he never expected.

Time passed and he became a friend of this new area. People who wanted something different and peaceful, they wanted to know that the world was better. Titan could only agree that things had changed, but he too did not know exactly how. He hoped that the torture would stop soon, and he felt like he needed to help. He felt like he needed to still find his sister at any cost. The feeling of levitation only increased his energy to do something in the world that was not his own. And at the same time recuperating with knowledge and strength, realizing just how encompassing the evil had become in this hyperactive land.

Titan had found some of the men in a small glade meditating. The top four men of this small quiet tribe were always patrolling their border even if it was to listen in on their surroundings. Their names were Michael, Lucas, Ignacio, and the old man Martin who first found Titan. Titan's stroll had been quiet and he took a minute to gaze at these men. He wondered if they had heard him but figured he would walk as close as possible before alerting them. 'Probably wouldn't be a good idea to scare them'

Titan mused as he crept on a slight breeze. It had taken awhile to close in on them and he stood for a moment to smell the air.

"You are quiet for these old men, but my ears still work." Martin said as he stood, while the other men whipped eyes around to see the tall Titan standing over them.

"The only other animal to sneak up on us would be a great panther, I see you always have your sword."

"Yes, because I will never be captured again." Titan explained.

Martin moved to grab Rafael's staff to give to Titan. "Then we should teach you how to use it."

A quick thrust took Titan's wind as the staff was placed quickly in his stomach by the old man. Titan took a second to regain before swinging the staff only to have it knocked out of his hand. Angry glaring eyes located the staff as Titan would not be the one to give up anytime soon.
"When you kill somebody with a sword you must be fast, move through them and keep going because you do not want to think about what you have done."

"I have already seen what this sword can do." Titan said at the thought of unsheathing it but that thought was quickly dismissed as he remembered the bloody mess of what he had done.

Calm ran through him before making his next move towards the old man. A ruse had been intercepted as vicious wood collided in vigorous bursts. A steady stream of offensive and defensive blows cut through the air in small cracking bursts. KLACK, KLACK … KLACK, KLACK and air as a deft motion had been made in the blink of an eye. Titan's staff had been placed close to him as a ducking motion that put him directly at the old mans back, staff poking onto his flesh. A brief moment passed before laughter made both men move from their awkward position. Laughter and clapping as the other men came to congratulate the new talent. Martin was the first to speak.

"You're lucky on that one but it's these guys that you should really be practicing with."

Soon they were all verbally active as they gave up their secrets in the physical arts. This combined with active competition made for a much closer bond as they were all in this world for survival. Most of which had already been adverted but the real threat ever seemed closer to home. Titan found himself meditating with the others before and after their activities which made his demons fade, and made his focus and speed even more incredible. This tended to make everybody take another step in progress as their competition mounted in a sided tournament. Two who started to become jealous of Titan both attacked him as they had been beaten in the same day. He fought both for awhile but they took hits on his arms and legs until Martin began to help Titan. It did not take long before all were fighting to help or attack. The fight was brutal and to Titan it seemed like it lasted an hour but the sun still sat on the horizon with gleaming light burning through red clouds. They were all welted with cuts and bleeding faces but they felt good and ready.

Titan and Penelope also felt oppressed and different-unwanted.

They made plans to leave.

The fire

The last thing he remembered was driving in the lead of their entourage...they had passed the army-trucks, tanks, small fast dune buggies equipped with machine guns... these were her core army- her massive deadly gang. They all died so quickly.

The rail burst tossed tanks like ants in the wind and destruction so great and fast that Marcus was amazed to even be alive they were so close. And he laughed hysterically as he drove and a brilliant light rose in the sky. He was maniacal and dreamily looking up while the entourage had surrounded him and Angelina drove a needle into his neck. The light turned off and there was nothing.

*

Marcus woke to the smell of sulfur. He couldn't move as he realized that he was strapped down to the same table that held Titan. He struggled as the old doctor moved the machine out of its room. The old man moved with a dead systematic pace. And as the machine took its place over Marcus's hand he screamed in anger before the fire was placed in his veins. His muscles twitched in pain as the burning poison found the nerves that made his eyes bulge. He tried to relax and meditate as they had taught him in Terra. As he imagined fields of green and flowing blue streams a burning fire broke in and devastated the scenery. The burning liquid destroyed and evaporated the stream, and he screamed in pain and anguish. He breathed deep- panted as the machine gave him a break while the doctor moved it to the other side of the table. Meditation did not seem like it was working as his hand felt like it was melting throwing painful spasms down to his torso. The first injection made tears spring out of his bulging eyes. He rolled his eyes back and tried to focus on the pain itself, in order to subdue the burning. To accept his position was sometimes the best way to let go. It did not work. Soon he was placing himself on the meditative path as the doctor pushed the machine down to his feet. He imagined his feet were placed in burning, flowing sand as a fire leapt through his legs. He moved through this flowing sand as a shadow crept out of the scorching waves. An image appeared as his lost love Katie smiled which made his heart pound. The image was clarified as she hugged and kissed Marcus. His knees wanted to crumple as a cooling sensation swept through his body. He could not find words to put in his mouth as his mind was flooded by emotion. Grief overcame him as wind put chills through his body. His lost love kept smiling as she kissed his cheeks and spoke into his ear.

"You will be forgiven if you fight for what is right. And I will be with you… forever."

She kissed him with a rush of cold and began to back up and fade as Marcus reached out for his lost love. Swirling heat crossed his mind as his grief turned into anger. He could not take any more loss. Muscles bulged as he twisted while the doctor moved to the other foot. He could not feel the numb cold hands and he wondered if they were still there. He rolled back his eyes to stare at the evil, dark man. His body twisted as his arms bulged with lean sinew. The straps held as the table flexed and broke in a snapping sound of plastic and metal. The old man was paralyzed in shock as Marcus unbuckled his arms, pulled his fingers out of the tough elastic bands. The doctor realized what he needed to do and he moved to the shiny toolbox which held the various needles. Marcus moved in a flash to undo the straps that held his feet. Throbbing limbs leapt from the table as the doctor turned with a needle in hand. Marcus was not seen until he picked up the machine which rose over the doctors head. The old mans eyes bulged which was the first emotion that Marcus had seen from the doctor. The heavy machine landed on the man's head in a skull crushing crack. Marcus then moved quickly as he knew that this room was being watched by surveillance. He ripped open the machine and grabbed the container which held the liquid fire. Needles were found in the toolbox and filled with the yellowish liquid. Pulling drawers open was met with a pistol and clips. People would die if needed and in a flash he was in the hall with a rolling toolbox in front of him. He was met by the expected resistance-bullets landed in and around the metal box. The first two people were shot they did not expect return fire while the rolling toolbox was pushed past them. Screams were heard by the rest as they were overwhelmed by the rolling thunder- they spent bullets for no reason and tried to reload their weaponry. Some pushed against it as Marcus had crept to the side to stab a man in his stomach. He counted three more that were slow compared with the furious Marcus. Two more were stabbed by the needles as the men fumbled with their weapons. Crippling screams echoed through the hall as the people had balled up on the ground. Screams that filled Marcus with hatred as he walked slowly up to the last man. The man dropped his weapon as Marcus landed his fist through the man's chin, whipping the neck around in a terrible snap of the neck. He then took his time as he cleared the jammed gun and picked up the remaining weapons. He realized that he had been grazed on the arm as he strapped a holster to his leg while also stocking his pockets with ammo. Soon bullets started to land around him as he decided what else he was going to do. He was stuck in a building with one exit that he knew of beside the windows on the sides. It was a large warehouse style building that had three rooms and two hallways in a tee shape. The torture room was one of the two smaller rooms as he wondered what was in the bigger one. He already knew that the third room was the doctor's surveillance room, which didn't bother him as he had already crept backwards to escape the hail of gun fire, and was now looking at the large double doors which seemed to hide some dark secret. He now remembered that there was a garage door on the back side as the terrain had sloped downhill to allow this. Hidden mostly by brush and how the building was made. He thought this as he unloaded a clip into the lock of the door. It collapsed easily as he kicked the door in and could not believe what he saw. Marcus realized that the safest place to hide something was in the last place that people wanted to be. He tried to take in the exquisite curves of the 2050 Lamborghini Diablo. He wanted to take this car but his goal was to survive. His eyes drifted to the other car which looked as though it was heavily armored. He wondered what that crazy woman was thinking as she put an exquisite ride next to something so dull and ugly looking. Something that looked like it had already been used to get away from some kind of skirmish or two. It looked old and probably was made around the same time as the exquisite car sitting next to it. It should be easy to hotwire he thought as he opened the heavy door and sat down in the hard old seat. Marcus thought that the hotwiring should be easy as he remembered his wiring classes from past events. Soon to be amused, his eyes locked onto the shining key protruding out of the ignition switch, he turned it, and was relieved to hear the dull roar of a nearly extinct high power gasoline engine. As he looked around at the controls he noticed what must have been the remote to open the garage door. He pressed the button, revved the engine, and squealed the tires as the door opened just enough to let the vehicle burst out of the garage bay. Marcus was immediately met by bullets ricocheting off of the exterior of the heavy skin. He was amazed at his good fortune with the armored vehicle as he sped to the back exit of the compound. Dirt churning from the powerful, twisting torque made a natural smoke screen as Marcus cruised out of the compound gates. At any other time this might have been impossible, but most of Angelina's people were already defected or dead.

Marcus had already thought of this as déjà vu as he sped around the narrow, dusty roads and out of the desolate town. Then as a powerful emotion gripped his soul and twisted his mind he turned in a quick 180 stop to muster himself and learn about the tools in his car. One button looked like the targeting so he pressed it to be pleased by the holographic windshield display. A fire switch lifted a gun out of the top of the vehicle while a joystick popped up beside him. For being old and out of date, this small tank was incredibly built. Marcus remembered one of the old James Bond movies that he had at one time seen with Titan among others. Marcus knew which side that he would take because the good guy always wins, even if he is going to die, which would be most welcome for Marcus, as he could not get his lost love out of his dreams.

In a flash of churning wheels and gunfire Marcus ripped through the compound of hate and fear. Men who had given up because Angelina was not there to witness her loves torture were now scrambling to put up weapons. People collapsed from the rain of bullets as Marcus twisted around the large houses and warehouse-style buildings. His mind burned as he knew these people for those who do not care about life which was the reason that he had to set up such an escort in the first place. His love was a casualty of war because people could not grow out of their destructive ways. The way that he had embraced, like the scum that was falling by this unfortunate and most painful turn of events. This destruction that he is giving will be well earned by these followers of Angelina's evil. Marcus targeted steadily and launched a rocket at the only tower he knew of which held heavier weaponry. His pursuers fired a missile before the tower went up in a burst of flames. The agile car easily dodged the incoming explosion as he barreled around the corner to the compounds opposite exit. The exit that was trying to close its gate until Marcus opened fire on those in the control booth. He cruised out of the compound as he thought of how Angelina had once bragged to him about how safe she was from the outside world. Marcus realized Angel was afraid of those people that she tortured and he was going to be the one to clean the world of its scum ridden nightmare.

The road home

Luna accepted her position as her new people thought of her as their most blessed, most revered, and her followers were steadily rising. Many people of all cultures and religions were looking for answers in this volatile, unforgiving world. They wanted a beacon of peace and prosperity and many were willing to die for the greater good. She already named a general after just a week and reports were around twenty thousand and growing for just her army what was once a religious factions militia, while the camp around the newly built structures were on the brink of pushing into the millions. Luna was in contact with New Terra as more people poured in from the swamped coastal regions. They were willing to give them food and supplies to build a new city in the area that Luna and her people claimed. She thought that this was strange that even the Terranian people were seeking shelter. She accepted this greatly and found she had ample supplies and resources to create what everybody wanted- a fresh start. Luna also knew that things would soon be more volatile as reports were that this demon from the south was still not dead. She only hoped that her force was stronger than whatever this warmonger could produce. Worry was always quelled by the unending work and cooperative help from all people building and growing. Structures were being planned with material that would appear out of the sand. It had only been a week since John's death, and so much had happened that Luna felt overwhelmed by the scope of it all.

She rarely took time to pray, but that night she wished that people would all just get along.

***

Titan and Penelope woke and she whispered to Titan,

"I can't be found because they won't leave me alone."

Titan turned his head and replied, "Don't worry I will help you- if you need to be rid of somebody."

They left with supplies that made them realize that they could. These people helped in the most incredible way- with the least amount of way possible.

And they moved through dangerous and habited forest.

A blade at Titans hip and a gun, and a blade at Penelope's but no other. And she could run like the wind.

She knew this land. She knew it far more than Titan and she used this to her advantage until. She spaced herself too far... though small campgrounds and other small trails she moved where Titan would have a hard time to follow if not prepared. It was a game to her. And she disappeared.

He broke through the brush to find a familiar clearing with Penelope on one side and a black crouching shadow on the other. Titan moved to place himself in the middle with burning legs. The beast and Titan locked eyes as it sat down in a rigid statuesque fashion with rippling sinew ready to pounce- ready for a meal. It seemed like the first time that Titan and the beast were free in a place with no boundaries, and he wondered how the panther would react.

The shadow of movement caught the eye and the panther was gone while it seemed as though the sound of the forest crept back into tuned ears without knowing it. The girl hugged him from behind in a great squeeze that took his breath away. He had to pull her away and turn her before he relaxed and spoke.

"Let's go kill these people who want to harm you."

A small smile crossed her face and she spoke of this man who first raped her when she was –twelve- just because he could- even when many people knew.

They made plans throughout the rest of the day as they walked towards their destination. His plan was to creep in like a ghost to kill without a sound before using firepower. And after walking through the night till dawn they decided it was close enough that they should just get it over with.

Titan knew that most of this time he had been traveling further into Mexico and he did not know where he started. Captured, knocked out, taken, and then taken again for days until he was in some shitty compound- only to get ultimately used for bait. Some strange stupid capture just so people could get some laughs before whatever they had planned. He wondered how many techniques, how many different ways people could mess with somebody when he was so quick in Terra. It annoyed him even though he escaped.

Time passed and they made it to their destination.

***

Marcus knew that Angelina would not stop her war against the world until she was dead. He tried to travel fast because he knew that she would be gathering her underground armies.

She did not care how many she would lose even if it was a massacre. Although her losses were great she did not want to loose her power at any cost, and many would still join their warlord as they had always done even with dissent in mind.

Marcus was thinking of survival at the moment, and would stay out of reach as he had a long way to go before people would listen to him. The armored car was built well with a long range capacity but even this hybrid would need to refuel eventually. The only resistance he felt was when he stopped off the highway to refuel his car. He looked at the sign which read:

CLEAN GREEN FUEL - $11.95 PER LITER

Marcus did not have the money or the patience to wave a gun around a bullet proof box so he shot a missile at in instead. The speaker came on outside his window and blared.

"YOU WILL HAVE TO DO BETTER THAN THAT!"

Marcus replied, "I CAN AND WILL IF YOU DON'T LET ME FUEL UP."

The store attendant took a second and said. "OK, PLEASE- JUST TAKE AND DON'T KILL ANYBODY."

Marcus replied, "I don't want to kill, I just need the

fuel."

The green light came on above the pump terminal and Marcus jumped out to do this quickly before there was any more resistance. The fueling took a minute while he wondered how long it had been since a robbery such as this. The world had reinforced the safety of gas stations in the mid 21st century as a great shortage had struck the worlds energy supplies. Attendants were normally locked in bullet proof boxes to monitor each precious drop for the duration of their shift, but the main reason they were hired was for a first response security system. Marcus was not used to this system as he grew up in Terra where ninety-three percent of the cars were completely electric. The pump stopped and he jumped in the car as police sirens were heard in the distance.

The speaker came on with the attendant's voice, "SORRY, I CAN'T LOOSE MY JOB."

Marcus took the time to reply. "It's ok I understand."

He started the engine and floored the gas pedal to rocket the agile tank back to the highway. His destination was halfway met as he hoped to find Katie's father who was a Major in the military. Police cars were trying to surround him but he was not in the mood as he targeted the empty road behind him. He shot his last missile and as the ground exploded a police car fell into the new pit and flipped. The cars on his sides fell back as his journey would not be stopped by these people, who were most likely police in bed with corruption and Angelina's terrible ways. Marcus smiled at the plume of dust in the rear view camera and felt as though things were turning out ok. He was close to the border with little resistance and it would only be another day before he reached his destination.

***

Titan and Penelope traveled far throughout the night, and her old home was close as she told Titan of whom she hated and why. People had tortured her throughout her life as she wasn't seen by most as anything until she blossomed into womanhood. Then people would take advantage of her in any way possible. Her own family was the first to exploit her as they were heading to her own uncle and aunt's house. Titan was fine with helping as he wanted this girl to feel as free as anybody should be. Quickly and quietly they had crept through her old childhood town without a sound from anybody. She let Titan know of their target before they crept to the back where she knew a window would be good access. Penelope opened it as if she had before and moved inside as quietly, as a falling leaf to open the door of the basement. It broke the silence with a squeak which made them move even faster. Dark shadows rushed through the twilight and into the room of the sleeping people. Titan could see two mounds on the bed and was amazed that the squeak did not alert them. Penelope turned on a light and cried out.

"Die, you demons!"

The man turned to witness the blade of her knife cut into his own chest . She pulled out the sliver of metal as a fountain of blood poured onto her and she struck again. By now the woman turned to witness this but she did not scream as Titan had expected. She dove for her nightstand to uncover a hidden gun. Titan's eyes were fixed onto her. As she swung the gun around, Titan moved, and in a flash of motion cut the arm from the woman. She cringed as blood gushed from the wound, but did not have the time to scream as Titan's next motion brought the thin steel through her neck in a whooshing arch. Penelope had finished stabbing when she realized that her uncle was dead and was caught in a shock as she witnessed her own handiwork. Titan moved to her side and noticed a calm look fall upon this young girl. Her eyes looked far away and he knew that she found some kind of peace out of the death of these people. She still gripped her bloody knife in two hands with its dripping blade pointing down. Titan hoped that she was still alright and could get over a traumatic event such as this. His internal question was answered a minute later as Penelope spoke in a beautiful tone.

"Are you ready...for the next house?"

Titan only nodded his answer as he looked in her re-born eyes which were glistening with serenity. She smiled at him before her motion for him to follow, and their trail grew into a string of bloody murders as they visited many evil people. The growing red of dawn did not even stop them as they found a vehicle in which to travel on their mission. A justified higher goal that they were both getting very used to, and very good at.

***

Marcus had driven long and far, with plans, and questions constantly crossing through his mind. The desolate land brought signs of many people on his journey, those who looked as though they were moving away from the coastal areas, and away from the ocean. When he had crossed the same latitude of Angelina's war he remembered Luna's face as he had pulled the knife from the large man. Such terrible anguish had fallen upon her as if she loved this man that he did not know. A similar feeling was felt when he thought of his lost love and hate for himself furrowed his brow when he thought of what he had done. Thoughts pressed onto mind and even his love Katie would not have been killed if he hadn't pushed her to move to his country. She was trying to get away from her Father and her old life. Her father was a hard military man, but a loving father that Marcus had once met, who looked at him with kind hatred as he knew that his daughter would be better in a prosperous city. This was nearly two years ago now and he wondered if her father would still be living in the same house in the middle of the broken down city. Mexico seemed nicer than many parts of the United States which had dwindled into so many slum states. Mexico had greater farmland with temperatures that were not as inconsistent which led to a stronger core middleclass throughout the age of 'Industrialized Technology'. It was unfortunate for the masses that most were pawns in the game for greed and money. And while great recessions struck the United States constantly with corporate and political corruptions, people were more willing to target the rich. The rich class which owned virtually everything was bombarded for awhile with arson and murder. Marcus now realized that many people had put their money elsewhere and now knew that it could have been New Terra among others. His calculating mind brought tears to his eyes as his past events came into focus. New Terra was greedy, but for a greater cause to save the world from that extraterrestrial menace and now knew that the survival of a moral right was at hand as this greedy nature on a small scale was even more destructive. He thought this as he passed a great movement of people all carrying as much as they could. This was next to a newly formed road which seemed as though supplies were being brought from New Terra. Marcus wondered what news he might have missed while unconscious in the torture room. Time moves strangely when drugged and you cannot see the sun- so what would he have missed? His questioning mind would have no answers as he rode through the last couple hundred miles with no quarrel. Entering the small town of Katie's father he realized how it changed from a slum into more of a military compound. Road blocks created strange shadows and motion behind streets that he could not see as he rode slowly through the bends of the unkempt area. He realized he was being baited in a maze and he put the car in reverse. The quiet was lifted by a trembling thunder as the next block passed, and he saw it too late. 'STUPID,' he thought. Two roaring tractors now plunged at him in a sandwich vice grip as two more tanks closed in on front and back. Seconds slowed down as crushing metal and glass popped and smashed around him. The two tank barrels gleaming down at him as he felt like a mouse caught in the beast's trap. Marcus raised his hands in defeat as the tractors pulled away enough to get his doors open. Marcus had to kick open his own tank door and exit with raised hands. Men immediately appeared and closed in to grab and disarm him. He only still had on the shorts, the gun strap, and a jacket full of supplies and guns that he had taken. He felt strange as he was bound and blinded. Always a prisoner of war, but as he was not handled too roughly, he thought that it might not be so bad.

The trip ended briefly as he was taken to what he believed was most likely the house of Katie's father - Tom, Tommy to his daughter. He had always pictured the old gangster movies when he thought of that name, but knew also that he was in the Army, a fact that he tried not to think about. The dark bag was removed from his head and he glanced around quickly at the people that were staring at him. Large dark-looming men all strapped with guns glaring down at him as if he were an ant. The room around him looked dark as one spoke in a demanding tone.

"Tom is in that room at the end of the hallway. Don't make us go in there."

They pushed him as they walked though the large room. Opened a door and thrust him into a room that seemed large like the rest.

The shadow of the man on one side of the room became apparent and he knew who it was. This man was big, not as big as that giant he now thought of, who he killed in a fit of rage. This rage was now fear as he wondered how he would take this. A head lowered slightly before the large shoulders turned. Eyes opened to glare at Marcus like fresh meat, with a face that was abnormally calm.

The man stood tall to ask, "Why did you think that she would be safer with you?"

Marcus answered quickly. "I never thought Katie would die, she wanted change and you attacked us, I never wanted her death. Katie was my life."

The last part changed both of their faces as Tom's turned to a grimace of rage and Marcus's face seemed to melt. Tom moved incredibly quick as he soon placed himself in front of Marcus with a crashing fist landing on his eye, followed by another. Marcus wondered how many he could take before he had to fight back, but it stopped after three. A swollen eye grew as he accepted the man's punches while a raging voice spat out orders.

"This is what you are going to do. One, tell us anything we want and maybe you will live. Two, never see me again or you might just get the same pounding. Now go see the general because he wants to speak with you."

Marcus left gratefully without a word and he let the men take his arms and bag his head. They didn't even bind him so he knew he wouldn't be harmed much more… hopefully. He also knew what he wanted to say. He needed their help for the sake of better times. Something he knew as peace that most took for granted. He wanted to feel that again someday, but knew that it would not come anytime soon.

They moved and he was loaded into a vehicle. It traveled fast and the vehicle stopped and Marcus was led into a building with a constant shoving and pushing down the halls. He felt like the hard times were just starting in a way. How had he been led into this position? 'Maybe by fate,' Marcus thought, as the last push made him catch his foot on something into a stumbling fall. The men laughed as they unbagged his head and told him to wait and the door closed and locked. It seemed like an interrogation room with no tables or chairs, Marcus noticed- with a drain in one corner. Chills ran down his spine as he realized that Angelina was not the only one who was capable and willing to give torture. The entrance had a ledge which had tripped him and he looked up too see the many sprinklers to wash away the evidence of activity that 'didn't happen'. He felt close to death, but now for some reason, did not want to die like before. He felt absolved of past crimes for some strange reason and felt like he wanted to right those wrongs.

The door's knob turned and was swung open to reveal men standing in the doorway. One man entered as the door seemed to shut on its own. The man glared at Marcus with eyes like a serpent's, or worse while a hand rested on the butt of a pistol. A moment passed as the man studied Marcus who could definitely be his final judge. The man grinned and began to speak in a sharp tone as Marcus noticed his military name patch- LARSON.

"I believe things do happen for a reason, and you might be just what we are looking for. Military factions are trying to gain power because of your country's mistakes. Are you willing to be lead into battle to fight for us- into New Terra if necessary?"

'This man doesn't want to kill me, he wants to use me like a tool to discard of later,' Marcus thought. He replied as he walked forward to extend his back and puff out his chest for effect. "Yes, I can. I need to help those who I have hurt, and I need to start with whatever you need."

The grin looked devilish from the General but Marcus did not show his sentiment as the General opened the door to welcome his new man from the room of death. It felt ironic, in a way, like a rebirth that Marcus did not want to be apart of. He could not show these feelings and suppressed them as strength took over mind and body. The hard times have just started…

***

A light world brought a dark figure into being and it looked like her father. A massive being- something so deeply wrought. It felt good to be in such presence. She wanted more. The image grew deep and dark and a man with a lined face crept into view. She woke. People moving around her and she realized she was in a war room and people were dying on video cameras.
She wanted to remember the dream that she had that night but it was not fully there.
The powers that be were at war and America was stretching it's might on the southern threat. Another gangland army was being demolished by something so powerful. A massacre- and Alluana felt sick to her stomach.

War

Titan woke to the smell of rot and blood. Light was coming in through the dirty windshield but it was still dark out. He was sitting in the driver seat of their stolen vehicle while a sleeping, bloody woman sat next to him. Tapping sounds rose around him and he got out of the car as rain beat the metal with small fingers. Rain was something he hadn't seen in some time. It grew and started to pummel him in sheets as it cooled and cleaned his body. He watched the girl get out of the car to stand next to him, his calm unhindered as lightning crashes around them. They did not move as they absorbed the last minutes of hard rain, light illuminating the shadows of clouds while the storm lifted from its long, weary path. They looked at each other with the same bright eyes and knew that things seemed to be…ok. They both moved back into the car and started the engine. Titan wondered what would be next as he looked at the used dirty blade while the girl spoke.

"We are close, just a little more south. It's early morning… so we should strike or wait?"

Titan thought of his sister's questions and how they always seemed oddly like a statement and a question. He noticed how thin she looked and wondered when they had last eaten. Just yesterday at a victims house, but it wasn't much. No stomach for food after killing people. Titan wondered how many he ranked as far as killers go. Thousands of aircraft, ships, and submarines had been destroyed by him and his crew eliminating the world's weaponry and soldiers. Titan realized he was a wanted man and wondered what price most would give for him. Very wanted he thought as people were always feeding on other peoples fear and turmoil. It didn't take long to answer.

"Strike."

Titan floored the car down the narrow path and out of the small, forested area that they parked to camp for the night. Penelope explained the terrain, the house, and where they should park as they sped down the desolate early morning road towards the city. Things seemed to move so quickly and so crazy all of a sudden as it came into view. Dark smoke rose into dark morning twilight. City streets were littered from what looked like an earlier war. He drove fast but it was closer than he expected, he took the last left from her commands and she abruptly cried out to stop. Titan turned off the car as they ducked down while a car crept out of the gate. The car rolled by without stopping and he noted that it was a limo. Titan thought of how they must be so rich next to the slums of everyone else for they had parked by the run-down cars like their own. Titan reloaded his pistols and grabbed the small submachine gun that they had taken on their murderous route. He knew that this could be very different as people were most likely alerted of these close murders. Penelope looked disappointed as if the people had left forever. They waited like statues in the growing light of day to see the limousine return to enter the gate which closed quickly. They got out of the car to walk to the corner of the ten foot wall and crept into its twilight shadow. Titan lifted the girl over the ledge and followed in a jumping roll over the wall. They were good at being quiet as they crept lightly to their bloody goal. They crept behind a row of bushes to see tired-looking people following a man wearing pajamas inside. Titan noticed her eyes get wide as they ducked down to withdraw themselves. A minute passed as a murderous meditation fell upon their minds. They glanced at their prey and pounced as the tired people walked slowly to do their jobs. Most were security or house keepers who were kept awake by a scared madman. A man who felt how past deeds had haunted him already and now felt like something was coming. The perfect time had come as death rushed into the door that was being closed by the last servant. Open eyes tumbled to the ground as the next two were already downed by the girl. Security turned to witness quick death as three people lay on the foyers floor. Titan moved in a lightning grace as motion slashed through the necks of awestruck victims. Six people had died before they entered the house with guns in hand. Fire rained down on those whose faces were white with fear. And they moved through the building to exterminate these people.

Titan counted eleven who were killed in that room alone, so that made seventeen already. He remembered that he counted thirty three the nights before, so that made fifty deaths in the last couple of days. Titan thought of how he would now be remembered as a murderer. Maybe even a righteous murderer, but always as murderer. It was probably over fifty thousand he thought as he crept around to a large dining room entrance. He looked to see two security guards running towards him as he took aim to fire. Fifty two and counting... Poor guys were tired and shouldn't have died for a rich asshole, Titan pondered economy and politics as he made his way through the dining hall and kitchen. Penelope joined him as they looked through the pantry for the best foods to eat, and they took their time to fill hungry bellies.

***

The image was so distorted. She wanted good things to happen and they –were- actually happening. The American army moved on- even helped so much. The new town the -new City- was so different in just so many days. And it was as if she was the center of it all and she never felt or did anything. They paraded her, and they worked so hard, and for so much, and she did nothing.
Eventually she wanted to go to her home and she

knew it was not the same so she wanted to go to Johns home.

She had to.

***

The morning felt like the same. The same smell of blood. The same dark feeling, but the light came in flashes this time. Lightning before the rain made the day seem new and different. Titan looked at Penelope who was waking. Rolling thunder echoed through the large valley that they were looking down at. A large valley with many buildings in a sprawling suburbia, and they were parked among others on a mountain overlook. Others, who were camping, some like hermits, were also waking for the day as the grumbling storm intensified. Titan tried to imagine what kind of distance they crossed in their bloody exile. They treated virtually everybody like enemies or at least the people surrounding their targets. Titan had a sense of accomplishment and of freedom. They used deadly offence to defend their safety and had killed many powerful people who deserved to die. Titan might have felt completely justified- if it hadn't been for the smell. Why did the smell of blood seem to linger? He tried to stay clean but the smell held onto the old clothes he was wearing. He decided they weren't over yet as dark clouds erupted in flashing thunder. He and Penelope turned to each other as it seemed like their life would be death. She looked away and towards the beautiful animate sky as she spoke.

"It's not far, but he lives in a castle. This could be our last stop but he is a very evil man."

Titan started the car and did not take long to answer.

"Then... where do we go?"

Titan drove as usual as Penelope gave directions. The sprawling city was desolate and dark as they wove through the roads. Penelope could see people who looked at them as if death was here. The stories of mass killers were spreading faster than they were moving, and she realized it was one reason why they were still alive at all and not someone's prisoners. This man did not harm her but he did have friends who would- who had. She thought of those stories as she told the big killer what she knew. She loved Titan and would kill or die for him in a second. She looked at him and asked.

"Did you notice it yet?"

They had been moving along a wall and Titan realized it *was* like a castle, with walls reinforcing houses, barricading the rulers from the rest of the people. Titan wanted to get in and could not help but wonder how. He turned the corner to follow the wall while asking,

"How do we get in?

Another turn and the gate had already fallen. He stopped to watch people flee and was not amazed to find that the uprising had already begun. With all the dissent it seemed to make things a little bit too easy. This was not the same- Titan could feel something terrible, pulling the car forward very slowly the last few people were shot before the exit. He stopped, didn't know what to do- still had to live in order to accomplish this mission, and drove the car forward very slowly. Was met head-on by an entourage of black vehicles, and he put the car in reverse but he didn't move the car- he remembered the stupid Reese from his first terrible kill. The kid would never back down and he always moved forward. The black limos were going to drive around them. Titan slammed the gears in drive and slammed into the first car, quickly he threw a grenade under the vehicles while Penelope followed suit. Men exited the vehicles and guns waved over Titan and Penelope who ducked back into their car. They were oblivious to the grenades which had been placed perfectly under the three black cars. It seemed like a second passed slowly as the man screamed for them to exit. Titan and Penelope grabbed the back of their heads as bullets ricocheted through the glass windshield- before the explosions pummeled the earth as two drums throwing cars off ground, and men to the hard pavement. Small secondary bursts of flame were ignited by fuel tanks on the vehicles. Screams followed as hardened men now felt for mortal wounds. Titan exited from his shaken car, and shot the men in the first car, then the downed screaming man, and picked up his automatic weapon. Penelope opened fire and Titan turned his sights on the men who were exiting the fourth car still within the gate- a stretched limousine. Men still dressed in garb of authoritarian power fell in a hail of bullets. Titan was cautious as he crept around the wall to enter the large courtyard. Walking over dead bodies as he circled the limousine to find a gun pointed at Penelope's head. The man was still sitting in the back of the limo, and Titan could not see him, but the man told him to put the gun down. Titan obeyed as he looked into Penelope's large eyes. She seemed calm as though she was ready to die. Titan would not allow that to happen as he felt the weight of his sword hanging on his belt from its old makeshift scabbard. She blinked at him and did a backwards roll behind the opened door, but the man was quick as he lowered the gun and shot into her leg. Titan's movement was built into one fluid motion as his arm made a arching crescent as he grabbed the sword, stepped forward, and swung through the man's arm. Screams of new found agony bellowed out of the man as Titan pulled him out of the car and threw him to the ground. Titan sliced the other hand in half to make sure the man would not do harm and he turned his focus to aid

Penelope. She was sitting up with blood pouring out of her leg, and Titan was fast as he tore off his shirt to make a tourniquet. She looked at him with big loving eyes as her face turned a ghostly white. He kissed her forehead and told her that she would be alright. She smiled and spoke into his ear.

"I love you."

He picked her up and walked away from the screaming man and out of the walls of the enclosed rule. Exiting the gate to stop before a growing crowd of people- they seemed to appear out of nowhere. Titan did not know how they would react but found a growing cheer for him and the girl in his arms. Smiles grew on the peoples faces as many stormed the castle walls to claim their independence from tyranny. Good people also noticed the dire need for medical attention as they told Titan to follow. An ambulance was waiting past the crowd at the end of a street which made Titan's heart sink as he felt like everything could be alright. He would not leave Penelope and he cradled her head as medics administered the painkillers. The doors closed and he knew that a willing doctor would be more than happy to help as the ambulance began to move at its steady pace- lights flashing in the dark morning. Relief flowed through shot nerves as he felt this generosity and looking down at the bloody scabbard holding the blade he hoped that all would be peaceful for some time.

***

General Larson could not stop thinking about the plans that he wanted accomplished now. He and other ranking generals were sitting around the large room getting ready for their meeting with idle conversation. They were the new government as the infrastructure collapse did not stop at America's borders. Politicians were in hiding or already dead as the uprising had ousted the authoritarian leaders who held power for years. The unyielding statues of bigots were now pillars of crumbling sand as the years had undone the ties of organized politics. This had only created militias- military states with their own aspirations for the world. Even these military governments had hidden from the power of New Terra's weaponry, but now information had arrived which put their crosshairs over the country that they feared. Their intelligence had confirmed that no weapon-craft satellites were patrolling the skies. The time to act was now if they were going to capture the country which had made such devastating weaponry. They wanted to capture the gold which was hidden in the dragon's lair- this power was seen as high treasure in their greedy eyes. The highest ranking man entered, General Burnett with his entourage of personnel. This was the man who initiated the black government with their military rule. Larson imagined him as the one obstacle before his own leap to the throne. It would be the most difficult thing to bring down his own friend among the following of his peers but he knew that anything was possible. He pondered great acquisitions as the General began to speak.

"WE need to be focused in these times. WE need to be the strong. The ties between the four regions have never been stronger. Now it seems that a new State is emerging and I don't care if it's over Saviors, Sermons or visitors from Mars, but it seems like the right time to take this southern region."

Larson's hopes to go to war with a new enemy and a new country seemed to be there. He already knew an exile who defected from his land and his Marcus could be his perfect spy to infiltrate and defeat this perfect enemy. So small this country, but with power like a disease that brings irritation and death. This enemy has not seen an angry American force rise like a tide in the distance. And his rising irritation fell on those last few words, as if the General had read his mind.

"WE will take New Terra as the time seems right. They fucked with the wrong land. Gather the troops but do it with the new technology we give you and let's do this quietly. We strike in five days, so move quickly and I will see you on the other side. "

General Larson could not be happier. War with New Terra will make war interesting again. He wondered what kind of weaponry both sides would surface man to man. Larson was already thinking of taking all of Mexico for just a moment. He almost wondered where the thought had come from as he walked out of the room gleaming with delight. These people would soon praise him and then he would start another war as these things could fall into place perfectly. 'Focus' Larson thought, just like the General said- one thing at a time. Now they have American technology which will surface to conquer the parasites… the shoe dropped on a bug. Larson could not help but think of death- he wanted to kill.

***

Titan did not know what to do sitting in the hospital as he also slept in the room close to the wounded girl. The medical crew seemed adept which allowed Titan to relax and sleep in the comfort of a fresh bed, though it felt like just a few hours.

Too much to do with too much going on, and upon leaving the room the nurses told Titan to wait for Penelope. They said that the operation went well, and that she was still sedated.

That was an hour ago and Titan was getting restless. He did not even know this girl but they had already been through –too- much together. He cared for her, and she had said that she loved him. He thought this was strange, as he figured that she was only using him- Penelope was very good at showing no emotion at all, but he realized it was something that she survived upon. Titan could not wait any longer, and as he stood after hours he began to walk- and large group of people appeared at the far end of the corridor. He noticed that they all looked like doctors but he also knew that people were behind them with no scrubs on and a killing mind ran red. Halfway down the hall and he had already picked them apart in his mind. Quick motions would create a result that he never cared to look at. It was just a fact of nature he thought- only the strongest and quickest could survive. Ten more paces and movement through the crowd created angry doctors who were pushed out of the way. Titan relaxed, a closer view and some looked like people from the past then a cursing Samantha burst through and hugged him. Titan felt shock. It seemed like he would never get his old life back and he had already tried to forget so many things.

"How have you been?" he finally asked.

"Better and things seem to be good." Sam smiled and kissed him before she said, "You single-handedly defeated the Guatemalan Dictator who has ruled for thirty years."

Titan thought for a moment and knew that the man had been some kind of warlord but he did not think it would have been such a big deal. What really amazed him was how far south he had traveled. He troubled placing any dates on the past events because he had always been moving.

"I could not have done it without that girl, Penelope. She knew all those people who we killed."

"I know, Titan. We know that you have killed and can help you if you need it."

"You're with doctors from Terra?" Titan asked.

"Many are. We all are working throughout the world Titan, and things have changed."

Titan could agree on this and didn't want to admit that he hadn't thought of Samantha for some time-maybe it was too hard. He accepted a life of death and couldn't allow himself to feel for others. He was nearly ready to kill people who he could not even recognize as his own countrymen. Titan felt terrible and bowed his head as he asked,

"Is my sister still alive?"

Titan expected the worst but was relieved to hear that his sister was alright. Luna was not only alright but she was heading some new gathering to the north in attempts to calm the warlord's. He hadn't seen his beautiful sister in such a long time- he had fought so hard for her. He almost compared her to Penelope which really made him feel strange. 'Why were my emotions doing this,' he thought, 'why must it be so hard?'

Samantha began to calm him as she told him the details of what had happened.

"Your sister was the lucky one to be captured by religious zealots who are good people. She started a movement to the south created by followers of a kind man. A self proclaimed Savior helped your sister defeat a warlord."

Samantha paused and waited for reaction from Titan but there was little more than a look into her eyes as he was mostly still.
She began again by talking of how the religious uprising fell on an interstellar event which placed earth in the midst of massive forces. In the first small battle the man had been killed right before the wake of gravity moved through earth.

"People always look for answers in troubling times," she said.

Many words he didn't remember but one line stuck with him as he fell into thought, 'The desert already has a city which has Fifty Million people.' Titan thought that it was more than Terra ever had in the past which seemed so strange to him in a way and he wondered how his home was doing.

"How is Terra?"

"You wouldn't like it," Sam replied. "More than half of the country has already left. They think that the

Americans will eventually try to conquer our land."

"Things change but I will die before that happens," Titan looked at Samantha with gleaming eyes and didn't expect the response that she gave.

"Good because I want to go back too. I have a flight back to help the 'exchange' and I'm sure they will be glad to see you."

Glaring into steady eyes, they both knew that they would die for their small country. They loved each other because of it- unbelieving that anything would ever harm the place that had grown so powerful. He told Sam to wait as he visited Penelope sleeping in her room. He felt that it would not be the last time that he would see her. Now she would heal and rest while his duty was needed back home. He turned to leave and felt as though she was following because the next destination would be more deadly than any of his past- it would be his last destination. A dark shadow lay still behind his slowly beating heart.

***

They breached the country's perimeter without any resistance at all. The shock of the deserted city was just as profound as Larson's anger. They already set up a small base within the borders of New Terra. There were no shots fired as fury burned in Larson's eyes watching the progress through the soldiers video feeds. Empty streets only brought more empty streets as the city looked as though people had just vanished. Small streetlights illuminated a 3D interactive city map that soldiers stood around to procure their location.

Ghostly verdant buildings were surrounding and confining as curving multi-tiered pedestrian bridges spanned the skyscrapers connecting the city in an iron web.

The soldiers were making plans to scout in smaller groups of twenty. They would have to if they were going to make any progress into this foreign landscape. Their new objective was to find anyone who might still be living in New Terra. Higher ranking captains were grouping their men as orders and locations were being made around the cities maps. They were going to try and scour the first quadrant of the Metropolis. Larson cringed as the groups of soldiers started the patrol of the abandoned superstructures. The ruse was obviously there which was only waiting to be found. Larson stared at the many screens as he waited for the first trap to be sprung.

Six hours later the men started to get lazy and Larson's wandering eyes also slowed while the soldiers began to stop and take breaks. Green hued beams of sunlight pierced through towering structures in the sweltering midday. Dry gusts of heat were felt in the higher levels of the structures while soldiers relaxed even laughing with each other as they wet parched lips. They felt like they already won the city as thousands of soldiers felt greedy prospects in their newly found surroundings. Ten mega-structures were completely searched and taken without an ounce of resistance. General Larson relaxed for a moment as he reclined and placed his head back in his hands. He would only have a couple more minutes to watch as his temporary base was being moved into the first of these buildings. He didn't know if he liked taking this country so easily. 'Wars were meant to be won, not given' Larson thought as his frown deepened. His steadily resting mind brought unfocused daydreams which seemed to bring the shadows out of the woodwork. Shadows began to move in the light with a strange and sticklike glimmer. The glimmer resembled ghosts in the distance as the General strained his eyes. 'THE TRAP IS SPRUNG!' Larson stood and picked up the small transmitter to start his barking commands.

"YOU'RE ALL SURROUNDED. USE YOUR THERMALS. DEFENSIVE FORMATIONS, NOW!"

It was already too late as the video feeds began cutting out. Shouts and screams from the battlefield rang through Larson's mind like a song. A calming shiver crept through his legs and he sat down to watch the carnage of those strong soldiers. Soldiers who would not die without a fight made for a vicious pounding of destruction. Video feeds flashing and dying as hardened soldiers began spinning in a crazed retaliation from their certain demise. Screams of close combat brought Larson's eyes to a man in a dark suit holding onto a knife - blood covered the camera before the video feeds completely blackened out. Larson's own men began to falling back into Alpha Base. Larson had things to do and would have moved faster if it not been for the hard on that he had produced.

***

Penelope woke to the sound of music. She moved and winced in pain but sat upright through the burning sensations. She felt good and relaxed and even smiled as she pressed the red alert button. A nurse came quickly and knew exactly what Penelope needed. Another IV was attached with more pain reliever added. Food was also brought without a question being asked but these were not on Penelope's mind. She was calm enough to accept these before she asked,

"Where is Titan?"

The nurse looked at her and smiled, "He left for awhile, to go back to his country, I think."

Penelope's head fell to her pillow as she started to cry. The nurse came to her side.

"I'm sorry, honey, I'm sure he will come back for you. Do you need anything else at the moment?"

Penelope sniffled and replied, "More food, juice."

The nurse returned with more and was queried.

"When can I walk again?"

"It didn't hit the bone, so just a couple of weeks."

"Can I contact Titan?"

"Yes we have a number for you."

"I need it."

The nurse returned to see the girl standing, testing her leg for pressure and pain, but standing on it with a large grin on her face.

"Titan and I are lovers, and I need to go see him."

"Okay," was the only response as she left the room to call for help. They took a while and spoke of sedating her, or possible detainment as she still seemed delirious. This medical office did not want any trouble and they took their jobs very seriously.

A large man entered Penelope's room with a tranquilizer. He looked around for her before he looked under the bed. Standing with a curious brow he turned to see a small girl standing before him.

"Ah, there you are… you ready?" he said with a grin as he lifted the needle from his side.

"Yes" was Penelope's reply. One motion was met with another as a small piece of metal had crossed the man's throat. He fell as gushing spurts of red soaked clothes and streamed into wide gaping eyes.

"Now, for that number…"

*Thank you for the Kurt Vonnegut quote from Slaughterhouse- Five.

This was a hard book to write with hard things to deal with, it was never written very well- it was definitely a first time failure, and a learning experience.

I would like to finish the story of Alluana and Titan, Marcus and the others all have many more chapters in their lives, and however bleak the future may seem- there will always be a great chapter to write!

Thank you for reading!